1 $\frac{4.25}{N}$ Mc

D1600742

This Egyptian alabaster jar, inscribed 'eight and one-sixth hins,' illustrates the fine craftsmanship that sometimes adds aesthetic pleasure to the technical interest inherent in the study of ancient metrological evidence. British Museum photograph of exhibit 4569.

HISTORICAL METROLOGY

A new analysis of the archaeological and
the historical evidence relating to
weights and measures

BY

A. E. BERRIMAN, O.B.E.
M.I.Mech.E., F.R.Ae.S.

Illustrated with
65 photographs and
line drawings

GREENWOOD PRESS, PUBLISHERS
NEW YORK

Originally published in 1953
by, and Reprinted with the permission
of E. P. Dutton & Co., Inc.

First Greenwood Reprinting 1969

Library of Congress Catalogue Card Number 70-91753

SBN 8371-2424-7

PRINTED IN UNITED STATES OF AMERICA

TO DAVID

231768

SOCRATES ON MEASURE AND WEIGHT

ACCORDING TO PLATO IN THE TENTH BOOK OF
'THE REPUBLIC'

Immersion in water makes the straight seem bent; but reason, thus confused by false appearances, is beautifully restored by measuring, numbering, and weighing; these drive vague notions of greater or less or more or heavier right out of the minds of the surveyor, the computer, and the clerk of the scales. Surely it is the better part of thought that relies on measurement and calculation?

QC
83
B45
1969

3 3001 00594 4987

Contents

Illustrations

Preface

On 11th May 1951 the Departmental Committee on Weights and Measures Legislation published a report recommending the ultimate abolition of our present standards in favour of the complete adoption of the metric system, including the decimalization of the coinage. This is a matter that touches every one: industry, commerce, the home, all would be affected by such a change; and a heritage of great antiquity would disappear from our daily life. It is not my purpose here to discuss this recommendation, but to say something that has not been said before about our weights and measures themselves: it is their history and their relationship to others that were used elsewhere in the remote past that come within the scope of historical metrology.

This book begins with an introductory summary of the results of research on which I spent much of my spare time for ten years, and I acknowledge with gratitude the unrivalled facilities extended to me in the libraries of Oxford University. It was only by chance that I read anything about ancient measures, and my interest in their origins remained casual until I happened to notice that published lengths of the royal cubit could be expressed plausibly by 20·6265 in. As this number has geometric significance in terms of the inch it seemed to offer a clue worth investigation, and this further inquiry I was willing to make because for numbers I have the liking that should be inherent in an engineer and also because metrology lies at the root of all the sciences that are concerned with the interpretation of measurements: if a study of archaeological records could reveal an intelligible history, it seemed almost a duty to make the search.

Following the introduction is a section explanatory of my analytical apparatus and conventions in terminology; and then comes the evidence in detail arranged geographically, with the Far East, represented by India, China, and Russia, placed arbitrarily before Babylonia, Egypt, and Palestine. An analysis of the earliest monetary evidence precedes a section combining Greek and Roman metrology, but any further reference to the metrology of numismatics would be outside the scope of this book. French pre-metric metrology is followed by brief chronological notes on the history of the metric system, and the remainder of the main text is devoted to the historical aspects of the weights and measures of England and of those used in the United States of America. No chronological significance is implied by this arrangement, and even the geographical boundaries are blurred by overlapping provenances. Some notes that are not strictly relevant to the main theme will be found in appendices, but there is no place in which properly I could refer to the earliest mathematical texts: it must suffice, therefore, to remark that these should be studied in parallel with ancient metrology, not only as an aid to perspective but also to sharpen opinion on the right answer to a particular question: Was the Earth measured in remote antiquity?

A. E. B.

OXFORD, 1952.

I

Introduction

THIS inquiry reveals the oldest extant weights and measures linked by significant magnitude ratios to those current in historical times, and suggests that this ancient metrology was geodetic in origin. One aspect of this hypothesis associates a linear scale with the sexagesimal division of the Earth's circumference; another implies that a decimal fraction of the square on the Earth's radius was adopted as the area for a land unit. The third geodetic aspect is apparent in a standard of mass hypothetically based on the density of water, and a fourth link with nature emerges from certain weights that look as if they were made to express the density of gold.

It is only in the light of the metric system that the first and second of these coincidences are visible; they could not have been discovered in antiquity and there is no record of any traditional knowledge of such an origin. Certainly the Greeks were unaware that the circumference of the Earth measures 216000 Greek stades.

The Greek stade measures 600 Greek ft. and the best evidence of the length of the Greek foot is the platform of the Parthenon, which is 100 Greek ft. in width by 225 Greek ft. in length. It was measured in 1750 by Stuart and a century later by Penrose:

the mean of their measurements gives a Greek foot that corresponds in length to one-hundredth of the sexagesimal second of arc on any great circle of the Earth regarded as a sphere.*

This estimate of the Greek foot does not rest on these measurements alone; it is confirmed by measurements of the Roman foot

---

* The geodetic appearance of the Greek linear scale was noticed by Jomard in 1812 and by Watson in 1915; see XV (1). Hitherto it has seemed an isolated coincidence.
in isolated coincidence.

▼

made a century earlier by Greaves. The relevance of this reference to the Roman foot lies in the fact, attested by ancient writers, that the Romans used the Greek stade of 600 Greek ft. but rated it 625 Roman ft.: any lengths purporting to be those of the Greek and Roman feet must conform, therefore, to this (25/24) length ratio, and it is the need to apply this test to the measurements recorded by Stuart and Penrose that makes it necessary here to refer to those reported by Greaves.

He was Gresham Professor of Geometry and about to become Savilian Professor of Astronomy when, in 1639, he equipped himself with suitable instruments and went to Rome specifically in order to ascertain the length of the ancient Roman foot. In the Vatican gardens he was fortunate to find a monument exactly suited to his purpose: it had been made in the first century A.D. to commemorate a young architect named Statilius Aper who had died in his twenty-third year. This large and elaborate cinerarium is now in the municipal collection and the instruments of the deceased's profession that it portrays in relief include the Roman foot-measure that Greaves reported to contain ' 1944 such parts as the English foot contains 2000.' This makes the Roman foot exactly (24/25)ths of the Greek foot derived from the Parthenon.

The Roman foot was rated 16 digits, which makes the length of this digit (1/54)th of a metre: there are 100000 of them in the terrestrial minute of arc measuring 6000 Greek ft. and it is convenient to call this distance a geodetic mile. If the Greek foot had been (1/100)th of an inch longer, the proportional length of 6000 such feet would have been equal to the Admiralty sea mile of 6080 ft.

The Roman digit was equal to the Egyptian digit of which 20 made the remen; and Griffith, when Professor of Assyriology at Oxford in 1891, identified the diagonal of the square on the remen with the cubit that Herodotus called royal. This relationship implies that the square on the royal cubit is twice the area of the square on the remen and was of practical convenience in land measurement. In particular, it made the land-unit that the Romans called jugerum equal to the square on 96 royal cubits and this area was (5/8)ths of the English acre; that is to say, it measures 100 English sq. poles exactly when the royal cubit is defined as 20 in. and (5/8)ths of an inch.

In terms of the 12-in. foot the pole measures 16·5 ft. but formerly it measured 15 ft. in terms of a foot of 13·2 in.; or it can be regarded as a length of 10 cubits of 19·8 in., and I have identified this as a Sumerian cubit of 30 shusi in terms of the linear scale on one of Gudea's statues in the Louvre. Gudea was governor of Lagash in southern Babylonia about 2175 B.C.* and the statue shows him seated with a tablet on his lap: on the tablet is the ground plan of a temple, and the linear scale; of this I have a copy that has been checked by the authorities.

This Sumerian cubit was equal to (24/25)ths of the royal cubit, which means that 12 myriad † royal square cubits were virtually equal to 13 jugera. This relationship has historical significance because Herodotus remarks that Egyptian warriors were rewarded with 12 setats of land (setat being the Egyptian name for a land-unit of 1 myriad royal square cubits) and Bishop Epiphanius in the treatise on weights and measures that he wrote in A.D. 392 says that the Palestinian jugon of first-class land containing 8 myriad Palestinian square cubits was equal to 13 jugera. It is apparent, therefore, that the area ratio of the Palestinian to the royal square cubit was as 3 is to 2. Also, it can be shown that the Palestinian field of one myriad Palestinian square cubits was equal to (65/64)ths of the English acre.

The English acre is the most intriguing of ancient measures because it is virtually equal to a hypothetical geodetic acre defined as one-myriad-millionth of the square on the terrestrial radius: if both acres are expressed as squares, the difference between the lengths of their sides is less than 1 part in 1200. The geodetic acre can also be defined as measuring one myriad square cubits in terms of a hypothetical cubit equal to one-ten-millionth of the terrestrial radius, and it is convenient to call this cubit A: its former existence is as plausible (or as incredible) as a cubit derived from the sexagesimal division of the Earth's circumference.

---

* See XIV(4).

† Myriad = 10000. The evidence compels me often to name this number; it had more significance in antiquity than it has now.

Decapitated statue of Gudea, governor of Lagash *c.* 2175 B.C.: on his lap is the tablet with the graduated rule. This is one of several statues of Gudea in the Louvre.

If there is any validity in the hypothesis of a geodetic acre it would seem necessary to suppose that it was the prototype of the Palestinian field as well as of the English acre, and this would imply that the Palestinian cubit originated as cubit A. It is described by Epiphanius in the same way as Ezekiel describes the cubit of the altar, that is as a cubit and hand-breadth: the cubit of reference in this description is the royal cubit.

If the English acre is drawn as a circle it can be inscribed in a square representing the obsolete Scottish acre and if this is expressed as a circle it can be drawn in a square Irish acre. Add only 14 sq. yds. to the Irish acre and it is the area of a circle 100 yds. in diameter, which is the published size of the outer earthwork circle at Stonehenge. A square enclosing this circle would have an area of 1 myriad sq. yds. and this probably was the size of a Hindu land-unit called nivartana: in any case it is relevant to remark that the area of the swimming-bath found at Mohenjo-daro is 100 sq. yds.

Mohenjo-daro (meaning mound of the dead) is the site in the Indus Valley where excavations revealed for the first time the former existence in that region of a culture contemporary and comparable with those in early Mesopotamia and Egypt. Among the smaller finds is a piece of shell on which a linear scale is most carefully marked by fine saw-cuts: separated by five divisions on this scale there are two lines specially marked and the distance between them is exactly 2 Sumerian shusi; I call this length the Indus inch.

A multiple unit representing 25 of these Indus inches survived in the traditional yard that Akbar standardized when he came to the throne in 1556 and still survives in the 33-in. re-standardization of this yard by the Government of India in 1825. British interest in Akbar's yard arose from a desire to establish the size of a land-unit called biga: it was found to be equal to the square on a linear unit, called jarib, containing 60 of Akbar's yards, but it was not realized that this made the biga exactly equal to 100 English sq. poles and, therefore, to the Roman jugerum. Instead of perpetuating this interesting tradition, the authorities imported the English acre into Indian metrology and divided it into 100 parts called decimals.

The double-jugerum that the Romans called heredium survived as the arpent in the pre-metric metrology of France but its importance in historical metrology is quite overshadowed by the significance of the French livre.   In 1742 the Royal Society of London and the Royal Academy of Sciences at Paris collaborated in the direct comparison of the English and French metrological standards and the result showed the livre to be (108/100)ths of the present English standard pound.

Now there is, in the British Museum, a Babylonian stone weight that I call mina N because it is engraved with a cuneiform inscription certifying it to be a copy of one that Nebuchadnezzar the Second made to accord with an earlier weight owned by a former king named Shulgi who belonged to the third dynasty of Ur and reigned about 2100 B.C.; this inscription, therefore, is documentary proof of intention to preserve a standard of mass for at least 1500 years, but it has not previously been noticed that this standard was the livre.   The mass of mina N is 2 livres, and this makes it equal to 3 Roman librae.

Moreover, I find that a hypothetical talent of 60 livres would be equal in mass to a Greek cubic foot of water at maximum density. There is now no physical example of this talent, but it can be identified with the water-weight of the biblical measure of capacity called bath if the cubit used in the measurements of the 'molten sea' is interpreted as cubit A.   The tradition of this talent as a unit of measurement seems to have been preserved in modern times by the tun of wine rated 252 troy gall. which is equivalent to 36 Roman amphorae or 32 Greek c. ft.

In English historical metrology the present standard pound is associated with the 14-lb. stone in a system of mass known from medieval times as averdepois,* and it was in terms of this system that Edward III defined the mass of the sack of wool as 26 st., that is 364 lb.   This rating reflects the importance of England's export trade in wool for it was adapted to the 500-librae (= 360 lb.) standard used by the Florentine buyers.   This sack of 500 librae was an exact English standard in another system where it was rated 350 mercantile pounds or 5 fotmals;

---

* 'The best modern spelling is seventeenth-century averdepois; in any case *de* ought to be restored.'   *Oxford English Dictionary*.

fotmal being the name of a mass equal to 100 librae, with a rating of 70 mercantile pounds or 72 lb. averdepois.

A document of this period, devoted entirely to weights and measures, defines this sack of wool as equal in mass to one-sixth of a cartload of lead; this cartload, therefore, weighed 2000 livres or 30 fotmals and the tradition of this relationship and of the name fotmal both survive in the customary fodder of lead that now weighs 6 sacks (= 156 st.) in the averdepois scale.

Sometimes the mercantile pound (libra mercatoria) was confused with the pound averdepois: it was known as the pound of 25 shillings at a time when the pound of 12 ounces troy was called 20 shillingsweight, but its earlier significance was 2 tower marks or 16 oz. in this scale.

The tower pound of 12 tower oz. was so called because it was the standard of mass in use at the Tower of London Mint until Henry VIII (possibly on the advice of Cardinal Wolsey) abolished it in favour of its rival the troy pound. Probably, Ethelred the Unready was referring to a prototype tower pound in his law requiring those in charge of towns to stamp their weights according to the standard of his mint; his rating for it was 15 ores. The masses of the Roman libra, tower pound, troy pound, mercantile pound, and French livre were proportional respectively to 14, 15, 16, 20, and 21 such ores. The mass ratio of the pound averdepois to the mercantile pound was as 35 is to 36.

Its use at the mint associated the tower pound with the coinage but there was a much earlier association with bullion if I am correct in believing that this mass was the Euboic mina mentioned by Herodotus in his account of the Persian tribute. Having explained how Darius obtained his revenue by assessing the twenty satraps appointed to govern the provinces of his empire, Herodotus remarks that the tribute in gold was reckoned by the Euboic talent but the Babylonian talent of 70 Euboic minae was used to weigh the silver. A numerical discrepancy in the text that has come down has presented commentators with a problem in arithmetic as well as with one in metrology, but if my solution is accepted the Babylonian talent weighed 70 tower lb. (= 50 livres = 54 lb.) and the Euboic talent weighed 54 tower lb. representing the mass of 60 c. in. of gold.

This may seem to be a surprising statement but I find sufficient support for the hypothesis to give the name gold-mina to a mass equal to (9/10)ths of the tower pound. It is not so much the use of gold that is unexpected as the use of the cubic inch; but it was this, of course, that enabled me to recognize the density. Among several extant examples of this standard is an ancient Egyptian

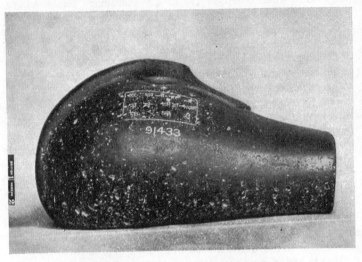

Merodach-Baladin's 30-mina swan-weight.
British Museum photo of exhibit 91433.

stone weight with a mass of 3 gold-minas; but its rating mark is 60, which implies a unit of mass equivalent to that of (1/20)th of a cubic inch of gold. Also in the British Museum is a stone weight, shaped like a swan,* that belonged to the Chaldean prince whose biblical name is Merodach-Baladin: it is inscribed '30 minas true. Palace of Eriba Marduk, King of Babylon' and its mass is equivalent to that of 48 c. in. of gold. The propor-

---

* This large example of a class usually called duck-weights has been identified as a swan by the Natural History Museum, and certainly the relatively royal symbolism of such a designation is more in keeping with the high ambitions of its owner. Some duck-weights represent geese.

tional mina can be rated 60 in terms of the mass of the well-known Persian gold coin called daric.

Perhaps the most remarkable evidence of the ancient use of the cubic inch of gold as a standard of mass is to be seen in a collection of copper ingots in the museum at Athens. Seventeen of them were found in the sea off the east coast of Euboea; their masses have been published in grams * and I find all these to be multiples of a mass equivalent to that of a cubic inch of gold reckoned as 315 gm. Another ingot in the same scale of mass was found at Mycenae; it is heavier than the others and weighs 75 gold-minas.

Among the smaller Egyptian weights are a few engraved with the hieroglyph of the word nub, meaning gold, and some of these have marked ratings in terms of fractions of the gold-mina. One of these gold-sign weights belonged to Amenhotep the First, son of Ahmose the Liberator: he was the second pharaoh in the eighteenth dynasty, which ruled Egypt after the expulsion of the Hyksos. Another weight in this group is inscribed 'The Gold-smith Homeri,' which suggests that the gold-sign may have been used as a certificate of accuracy by bankers and others who needed very accurate weights.

Some reference to the oldest weight in the world must be included in even the briefest summary of the evidence relating to ancient standards of mass: it is a pear-shaped stone (now in the Ashmolean Museum at Oxford) and I call it mina D because it has a cuneiform inscription signed by Dudu, high priest at Lagash when Entemena was governor there (about 2400 B.C.) during the third dynasty of Kish. The inscription has been translated to read 'One mina of wages in wool,' and the mass (which is a pound and a half) expresses this significant equation—24 English oz. are equal to 25 Roman oz.

Possibly it was as long ago as in Dudu's time that a craftsman living in the Indus Valley made the numerous small cubes of chert that were found during the excavations at Mohenjo-daro:

---

* 'Gram(me). There seems to be no possible objection to adopting the more convenient shorter form, except that the -me records the unimportant fact that the word came to us through French.' *A Dictionary of Modern English Usage*, by H. W. Fowler.

they are unmarked, but the frequency charts that I have con-structed from the published masses of 288 specimens show them to be weights intended to express fractions and multiples of the Roman ounce; they range from (1/16)th to 5 such ounces.

The names of Entemena and Dudu are both inscribed on a silver vase that is now in the Louvre; it is a wonderful example of Sumerian craftsmanship and probably was intended to con-form to some standard measure of capacity. Thureau-Dangin believed this to be a niggin of 10 qa mentioned in tablets of the Sargon period; if my interpretation of his measurement of the vase is correct, this Babylonian qa can be regarded as a prototype troy pint rated (1/60)th of the English cubic foot.

The proportional troy gallon would measure 230·4 c. in. and this may have been the original size of the gallon defined in the English medieval laws as a capacity for 8 troy lb. of wheat; but it became a wine gallon of 231 c. in. in the form of a cylinder 7 in. diameter and 6 in. deep. It is still in use as the standard gallon of the United States of America. In England this gallon had long been in use by the Excise when, in 1688, some busybody told the king that it was too large and that his revenue was being diminished thereby—the Guildhall gallon ought to be used, he was told, but the Attorney-General dismissed that standard because no statute referred to it and although he ad-mitted that he did not know how the Excise gallon had come into existence he was quite firm in his opinion that the Commissioners must not depart from its customary usage. Later, however, its validity was challenged again, but this time from the opposite side. An importer, prosecuted for refusal to pay balance of duty, said he had declared his wine in terms of a larger gallon of the Exchequer and that this was the proper standard for the purpose. The authorities dropped the case, having decided that the status of the Excise gallon must be strengthened by legis-lation.

The Exchequer gallon in this trilling was not the Winchester corn gallon (three examples of which were then in the Exchequer) but a calculated quantity reckoned as 4 quarts in terms of another Elizabethan measure dated 1601, which the prosecution correctly asserted to be the legal standard for ale and beer but not for wine.

A pint dated 1602 and evidently intended to be proportional to the ale quart was found by Commissioners appointed in 1818 to contain 'exactly 20 ounces of water' and on that evidence they recommended the standardization of the 10-lb. gallon that we use to-day.

This story of the confusion of the English gallons nevertheless reflects the respect for tradition exhibited by the evidence relating to ancient metrology as a whole. It is a picture presented here for the first time and I hope its new view of the remote past will enable this branch of archaeology to become a valuable source of information in a wider field of rapidly expanding general interest.

I(2) NOMENCLATURE

Monuments and rubble constitute the physical material of ancient metrology and individual specimens diminish in size from the Great Pyramid to fragments found in rubbish heaps that once were the homes of citizens in former civilizations. The Indus inch, for example, identifies here a linear unit engraved on a small piece of shell found during archaeological excavations in the Indus Valley. Mina D and mina N are terms by which I identify the standards of mass represented by two ancient stone weights, but in some cases the context requires that these names be interpreted as references to the weights themselves. In a slightly different category are such titles as Sumerian cubit and Assyrian cubit: I have applied the former to a length derived from a linear scale on one of Gudea's statues, and he was a Sumerian king; the latter relates to a unit derived from measurements that Oppert made at Babylon. In the arrangement of the evidence I have included both in the Babylonian section, and I disclaim any intention to imply by the use of such names any greater racial or chronological significance than is apparent on the surface: my concern here is exclusively with the metrological import of the facts.

As the main purpose of this inquiry has been to trace origins, I have used labels such as Roman foot and libra for earlier examples of these magnitudes wherever I have been able to find them, and there are certain words that I use uniformly for specific magnitudes without prefixing the names of the countries with which such units are principally associated. Digit, for example, is used here as the name of a specific unit of length and, of course, inch and foot mean the English units when used without qualification. Similarly, pound and its abbreviation lb. stand for the pound averdepois; libra and livre for mass standards commonly associated with Rome and France respectively and here used for specific magnitudes. Mass is a property of matter, and weights that balance each other have equal amounts of it: * I use the word mass when referring to the magnitude of this property in a weight and usually define it in terms of the grain (symbolized by gt.) or the gram (symbolized by gm.).

Water-weight and wheat-weight are terms used when referring to the mass of water or wheat that a measure of capacity will hold.

Relative magnitudes are expressed, for example, thus: (pound/ ounce) mass ratio = 16. This shows at a glance that the mass of the pound is 16 times the mass of the ounce; or, by multiplying across, the statement can be made to read: 1 pound = 16 ounces. The need for this unambiguous method is more apparent, of course, in such a case as: (setat/jugerum) area ratio = (13/12) approximately. These expressions are defined as length, area, or mass ratios on principle; although the context will have indicated that the setat, for example, is a land-unit and, therefore, a measure of area.

Mathematical symbols are used with their customary meanings; those appearing frequently include:

π = (circumference/diameter) = (area/radius^2) in a circle.
$\sqrt{2}$ = the square root of 2; $\therefore$ $\sqrt{2} \times \sqrt{2} = 2$.
$A^2 = A \times A$ = the square on A, in a geometric context.
lb./c. in. = pounds per cubic inch (density).

Generally, fractional numbers are expressed in the form, for

---

* The mass of a weight is constant and its usual symbol is m, but the weight of a mass (weighed by a spring balance) is expressed by the product mg and this is variable: it increases from 978 dynes per gram at the Equator to 983 at the Poles, but this difference is too small to be observed by ordinary means.

example, (3/2) instead of 1·5.   Combinations of such expressions include these examples:

$$(63/64)40 \text{ means sixty-three sixty-fourths of } 40.$$
$$(9/16)k \quad \text{,,} \quad \text{nine-sixteenths of } k = (9k/16).$$
$$1000(k/3) \quad \text{,,} \quad 1000 \text{ times } (k/3) = (1000k/3).$$
$$2(\pi/3)(30/2\pi)^3 \quad \text{.,,} \quad 2 \times (\pi/3) \times (30/2\pi)^3.$$
$$(9/10)\text{lb.} \quad \text{,,} \quad \text{nine-tenths of a pound.}$$

Usually expressions such as those given above are arranged so as to emphasize a particular aspect of the statement, and sometimes it has specific geometric meaning.   For example, $(1/2)(A\sqrt{2})$ means the semi-diagonal of the square on cubit A.

In the few places where the connection between a whole number and a fraction is addition, the sign $+$ is used.   Thus, $39 + (3/8)$ means 39 and three-eighths.   This retains the legibility of full-sized figures and expresses the metrological significance of the fraction better than does 39·375.

I(3) The Numbers k, π, and $\sqrt{2}$

Much of the analysis that has produced the results in this report has consisted in the direct comparison of numbers in a search for significant ratios.   When, for example, the mass of one weight is found to be twice or three times that of another the metrological relationship is obvious, but a ratio that looks like 1·3 would be dismissed as uninteresting by any one unfamiliar with the possibility that it might be intended to express the sexagesimal * number 1·296 that I have symbolized by the letter k throughout this text.

$$k = (6^4/10^3) = 1·296$$

No surprise should be caused by the presence of sexagesimal ratios in the results of an analysis of ancient metrological evidence, for the antiquity of the sexagesimal system of numeration is traditional; moreover, its application is established by the metrological content of texts inscribed in cuneiform on many clay tablets.   The particular importance of sexagesimal ratios to the

---

* Small numbers, appropriate to the rating of metrological magnitudes, are here called sexagesimal if they can be derived readily from powers of 60.

present inquiry, however, lies in the fact that some of the most significant aspects of the antiquity of the standards we still use are expressed by them, and I find the use of k a convenient convention that keeps this relationship in sight.

A peculiarity about the number k is that it can be expressed with negligible error by the ratio $(35/27) = 1\cdot29\dot{6}$ (decimal fraction recurring); such ratios exist in ancient metrology, but their sexagesimal significance would not be apparent if their connection with k was unknown. A further peculiarity of equal practical importance is the fact that the product of these two versions of k expresses, without increased error, the value $k^2 = (42/25) = 1\cdot68$.

Incidentally, the fifth power of k can be expressed with negligible error by $(128/35)$, and even this has an important application in the analysis of the evidence.

There is also a geometric aspect of k in the fact that there are k million sexagesimal seconds in a circle and, as the geodetic hypothesis is based on the concept of a spherical Earth, it is of some importance to examine the metrological evidence in the light of this geometry: for this purpose, it is necessary to use rational values for π and $\sqrt{2}$.

The quadrant (chord/arc) length ratio is $(9/10)$ if $(22/7)$ is used for π and $(99/70)$ for $\sqrt{2}$ in $(2\sqrt{2}/\pi)$. Both values were known in antiquity.

The numbers π and $\sqrt{2}$ express respectively the geometric length ratio of the circumference of a circle to its diameter, and that of the quadrant chord to the radius: it is apparent, therefore, that the quadrant (chord/arc) length ratio expressed by $(2\sqrt{2})/\pi$ becomes $(9/10)$ if $(22/7)$ is used for π, and $(99/70)$ for $\sqrt{2}$. There is reason to suppose that the geometers of remote antiquity believed this $(9/10)$ relationship to be a property of the circle.

Accepting $(2\sqrt{2})/\pi = (9/10)$ as an equation, leads to $(20\sqrt{2}) = 9\pi$ and to $\pi^2 = (800/81)$; all these values can be used with negligible error in the analysis of metrological evidence. There were other values, however, that were chosen for their convenience: for example, there is cuneiform evidence of the use of $(25/8)$ for π, and there are Egyptian texts instructing the pupil to calculate the area of a circle by squaring eight-ninths of its diameter. This is equivalent to writing $(1/2)\sqrt{\pi} = (8/9)$, or $(\pi/4) = (8/9)^2$; but it is interesting to note that if $(25/8) \times (8/9)^2 = (200/81)$ is assumed to represent $\pi(\pi/4) = (\pi^2/4)$, then $\pi^2 = (800/81)$ as accurately as before.

Perhaps the use of 31 for π^3 was known in antiquity; it is a close approximation.

With reference to the use of $(99/70)$ for $\sqrt{2}$: it is necessary to realize that $(140/99)$; $\sqrt{2}$; $(99/70)$ form a geometric progression, and that either extreme can be used as an approximation to the middle term.

I(4) THE SPHERICAL EARTH

The real Earth is not quite spherical; its equatorial exceeds its polar radius by 13 miles. These are the measurements made in 1906 by the United States Survey.

	Metres	Miles
Equatorial radius =	6378388 =	3963·339
Polar ,, =	6356909 =	3949·992
Mean ,, =	6367648 =	3956·665

This centrifugal distribution of mass causes the meridian length of 1 minute difference in latitude to increase from Equator to Pole in the manner shown diagrammatically on page 16: about latitude 50 it measures the Admiralty sea mile of 6080 feet.

In this analysis of ancient metrology I use the concept of a spherical Earth with a quadrant arc measuring 10 million metres: this was the rating assigned to the measured meridian quadrant that formed the basis of the metric system. This hypothetical spherical Earth, therefore, differs from the real Earth only in form.

Spherical Earth	Metres	Miles
Radius	= 6366200 =	3955·8
	= 10 million $(2/\pi)$ metres	
	= 10 million cubits A	

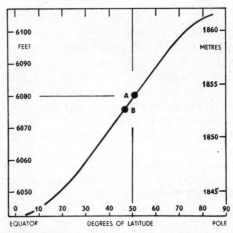

Diagram illustrating (roughly) the variation in the meridian length of one minute difference in latitude on the real Earth.

A = Admiralty sea mile = 6080 ft.
B = Geodetic mile = $(10^5/54)$ metres = 100000 digits
  = 1 minute of arc on any great circle of a hypothetical spherical earth measuring 10 million metres in its quadrant arc.

I(5) THE GEODETIC MILE

Two standards of reference are used here for comparison with the ancient linear evidence; one is the English inch, the other is the metre. In the metric system, as established in 1795, the traditional sexagesimal division of the circle was abandoned in favour of a measured meridian quadrant rated 10 million metres, but here I find it necessary to convert this rating into sexagesimal measure by employing the concept of a spherical Earth and defining a geodetic mile as the length of a sexagesimal minute of arc on any great circle. This makes the length of the geodetic mile $(1/54)$th of 100000 metres or, by calling $(1/54)$ metre a digit, the geodetic mile can be defined as 100000 digits.

Geodetic mile = 100000 digits, for digit = $(1/54)$ metre
  ,,   digit = $(1/54)$ metre = $(9/16)k$ = 0·729 in.
∴ (Metre/Inch) length ratio   = 39·366.

This geodetic digit was also the Roman digit, for 16 such digits make the Roman foot equal to $(3/4)k = 0.972$ English ft. and this is exactly the length of the Roman foot on Statilius Aper's monument, as recorded by Greaves in 1639. This example of the Roman foot was shown to be in accord with ancient tradition by the measurements of the Parthenon made by Stuart and by Penrose; these established its length ratio to the Greek foot as $(24/25)$, which is the relationship given by ancient writers. It is also clear from the evidence in general and from measurements of the Great Pyramid in particular that the Roman digit was a perpetuation of the Egyptian digit that was current in remote antiquity.

In Egyptian metrology there was a linear unit, of 20 digits, called remen and a distance of 500 remens became the stade used for itinerary measurements by the Greeks and by the Romans.

$\therefore$ Geodetic mile = 5000 Egyptian remens
= 10 stades (Greek and Roman)

$\therefore$ Circumference of the spherical Earth = 60^3 = 216000 stades.

No such traditional size was known to the Greeks. Eratosthenes of Cyrene, a younger contemporary of Archimedes, observing that at Syene at the summer solstice the sun cast no shadow from an upright pole, while at Alexandria (assumed to be on the same meridian) the length of the shadow implied that the rays of the sun were inclined to the vertical by $(1/50)$th of the circle, calculated the circumference to be $50 \times 5000 = 250000$ stades on the assumption that the distance between these two places was 5000 stades. This result was due to an accumulation of errors but even if he had found the circumference to be 216000 stades he would not have recognized it as a number of any special significance unless he was already familiar with the sexagesimal division of the circle in geometry.

These are the principal ancient linear units related to the remen of 20 digits:

Roman foot = $(4/5)$; cubit = $(6/5)$	Royal cubit = $\sqrt{2}$
Greek ,, = $(5/6)$; ,, = $(5/4)$	Talmudist = $(3/2)$
Assyrian ,, = $(8/9)$; ,, = $(4/3)$	Palestinian = $\sqrt{3}$

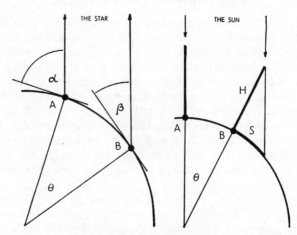

The circumference of the Earth can be estimated by measuring a meridian length from A to B, and the corresponding angle θ. In principle, this angle can be obtained, as $\theta = (\alpha - \beta)$, from the difference between the measured altitudes of a star; or, as $\tan \theta = (S/H)$, from the length of a shadow from a gnomon at B when there is no shadow at A.

I(6) The Radius of a Circle

In terms of its quadrant arc, the circumference of a circle is expressed by the number 4; and the length of its radius by the number $(2/\pi) = 0.636620$. Similarly, in terms of its sexagesimal second of arc, the circumference measures k million and the length of the radius is $(k \text{ million}/2\pi) = 206265$ sexagesimal seconds of arc.

Recognizing part of this figure sequence in lengths of the royal cubit published in inches, I thought that this cubit might have originated as the radius of a circle having its circumference rated sexagesimally in inches; and that, of course, implied special significance for the inch. This analysis supports the hypothesis that the royal cubit had geometric significance, and that one of its aspects is plausibly expressed by the rating $(50k/\pi) = 20.6265$ in.

It should be observed that this hypothesis implies great

antiquity for a sexagesimal scale in which the figure sequence corresponding to the fourth power of 6 (i.e. 1296) is significant. The traditional sexagesimal scale that has come down in association with the circle gives this sequence in its number of seconds: thus:

$$\text{Circle} = 360 \text{ degrees} = \quad 360 \times 60 = \quad 21600 \text{ minutes}$$
$$= 21600 \times 60 = 1296000 \text{ seconds}$$

As this number of seconds is one-tenth of the fourth power of 60, possibly the circumference was rated 60^4 units in remote antiquity: it would not alter the figure sequence.

I(7) Cubit A and the Geodetic Acre

The radius of the spherical Earth being 10 million $(2/\pi)$ metres, it is convenient to regard $(2/\pi)$ metre as a cubit of reference and I call it cubit A. Geometrically, it can be represented by the radius in a circle of 1 metre quadrant-arc; its length is 0·636620 metre = 25·064 in. As there are 1550 sq. in. in the square metre, the square on cubit A measures:

$$A^2 = (2/\pi)^2 50\pi^3 = 200\pi \text{ sq. in., for } \pi^3 = 31$$
$$= \text{area of a circle of } 10\sqrt{2} \text{ in. radius,}$$
$$\text{in which the quadrant-chord} = 20 \text{ in.}$$

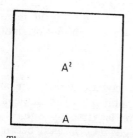 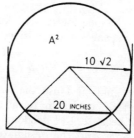

The square on cubit A and the circle of equal area. On the hypothesis in I(3), the length of the diameter can be rated 9π in., which is the length of the circumference in a circle of 9 in. diameter. For the geometric implications of these lengths in ancient Chinese metrology see III(1).

Possibly the origin of the inch was related to this geometry, for there is evidence in ancient Chinese metrology of a linear unit corresponding to the radius of this circle.

Evidence in support of the reality of cubit A is to be seen in the analysis of the measurements of the 'molten sea,' but this cubit seems to have been forced into rational relationship with the Roman foot and its square into rational relationship with the square on the royal cubit. From this latter association it acquired a length equal to $\sqrt{3}$ remen, and from its association with the Roman foot it acquired a sexagesimal rating in the form 2·16 Roman ft. = 64 cm.

The practical use of this sexagesimal ratio in Roman metrology is obscure, but the evidence of its existence is extensive. Of the six linear standards that Folkes measured on a tablet in the Capitol and reported to the Royal Society in 1736,* three evidently were intended to be proportional to the numbers 3 : 4 : 6 and to express 2·16, 2·88, and 4·32 Roman ft. Also, on Aper's monument there is (in addition to the foot measured by Greaves) a rod 96 cm. in length and this is exactly 3·24 Roman ft. in terms of a foot of 16 digits and a digit of (1/54) metre. Elsewhere also there is further evidence of the survival of linear units of 32 and 64 cm.

In this connection it is important to realize that the hypothetical existence of cubit A implies also that its original rating was in a form equivalent to $(2/\pi)$ metre, with a rational number in place of π. If (25/8) is used for π it makes the length of cubit A appear to be 0·64 metre, and this may have been the manner in which the sexagesimal multiples of the Roman foot originated.

The Geodetic Acre

In this analysis I use the name geodetic acre for an area of 1 myriad sq. cubits in terms of cubit A. The name acre is justified because the geodetic acre is virtually equal to the English acre, and the designation geodetic is used because, by definition, the geodetic acre is one-myriad-millionth of the square on the radius of the spherical Earth. Thus:

---

* *Phil. Trans.*, No. 442, p. 262 (1736).

$$R = \text{Radius of spherical Earth} = 10^7 A$$
$$\therefore \ R^2 = 10^{10} \times 10^4 A^2$$

$\therefore$ Geodetic acre = 1 myriad sq. cubits A = 4052·85 sq. metres

$$= 6281918 \text{ sq. in.}$$

2 million π sq. in	= 6283185	,,
English acre = 4840 × 1296 sq. in.	= 6272640	,,

The side of a square geodetic acre measures:

$$100A = (200/\pi) = 63\cdot662 \text{ metres} = 2506\cdot373 \text{ in.}$$

This exceeds the side (2504·5 in.) of a square English acre by less than 2 in.; that is by less than 1 part in 1200. A square of 2 million π sq. in. has a side of 2506·6 in., and a circle of this area has a radius of $1000\sqrt{2}$ in.; this, hypothetically, was the length of the ancient Chinese ying.

On the evidence of the measurements of the 'molten sea,' it seems safe to interpret the sacred cubit as cubit A. Ezekiel calls the sacred cubit of the altar a cubit and hand-breadth, and Epiphanius defines the Palestinian cubit of his day in similar terms: if, therefore, the Palestinian cubit originated as cubit A, then the original area of the Palestinian field must have been a geodetic acre and this appears also to have been the prototype that determined the size of the English acre.

I(8) The Inch and the Metre

The name digit is assigned here to the sexagesimal fraction of the terrestrial circle that measures $(1/54)$ metre, and its equivalent also has a sexagesimal rating in the form $(9/16)k = 0\cdot729$ in. This is a relationship of units that implies 39·366 or 39·375 as the (metre/inch) length ratio, for $k = 1\cdot296$ and $(35/27)$ respectively. In modern tables, this ratio is given as 39·37;[*] and its square is $1550 = 50\pi^3$, for $\pi^3 = 31$. This enables $(4/\pi^2)$ sq. metres to be converted into 200π sq. in., as the rating for A^2.

The royal cubit rated $(\pi/6)$ metre and $20 + (5/8)$ in. implies a metre of $39\cdot375 = 39 + (3/8) = (63/64)40$ in., if $(22/7)$ is used for π. Also, 2·16 Roman ft. = 0·64 metre can be interpreted as 25·2 in. in terms of this metre.

---

[*] In the U.S.A. this is the legal ratio.

I(9) GRAINS AND GRAMS

For convenience, I use the abbreviations: gt. = grains * and gm. = grams; also these sexagesimal relationships, which are correct (in terms of modern metric equivalents) to one part in 60000:

$$\text{(Grain/Gram) mass ratio} = (k/20) = 0 \cdot 0648$$
$$\therefore \qquad \text{Troy scruple} = 20 \text{ gt.} = k \text{ gm.}$$
$$\therefore \text{Pound averdepois} = (7k/20) \text{ kilogram.}$$
$$= 453 \cdot 6 \text{ gm.}$$

In English metrology the grain is the unit in which the legal standard pound averdepois has its rating of 7000 gt., and this is now the definition of the grain. Formerly, the troy pound was the legal standard of reference, and its mass was 5760 gt. in terms of the same grain. Earlier still, the tower pound was in use at the mint at the Tower of London; it was in $(15/16)$ mass ratio to the troy pound and, therefore, can be defined as 5400 gt. There was also a mercantile pound (symbol M) representing 2 tower marks = 16 oz. in the tower scale, or 15 oz. troy = $(36/35)$ lb. The relationships of these pounds to the Roman libra and to the French livre = $(2/3)$ libra can be tabulated thus:

Libra	Tower lb.	Troy lb.	Lb.	M	Livre	
14	: 15	: 16		: 20 :	21	ores
	27		: 35 :	36		mass ratios
			1	:	1·08	,, ,,
5040	5400	5760	7000	7200	7560	gt.
	350		453·6		490	gm.
			$(7k/20)$		$(7k^2/24)$	kilogram.

In the above formulae, $k = 1 \cdot 296$ and $k^2 = 1 \cdot 68$.

The pound (lb.) is k tower pounds, for $k = (35/27)$.

The arithmetical error in writing $(35/27) = 1 \cdot 296$ for $k = 1 \cdot 296$, is exactly $(1/4375)$. Applied to mass, this is equivalent to one-tenth of a grain in the ounce: for example, if the pound averdepois

---

* Formerly it was customary to speak of grains troy, and the symbol gt. reflects this derivation.

were exactly k lb. tower it would weigh 1·6 gt. less than 7000 gt. It is important also to note that k^2 can be expressed with an error of only $(1/4375)$ by writing $k^2 = 1·296 \times (35/27) = 1·68 = (42/25)$. Thus, for example:

$$100k \text{ gt.} = 129·6 \text{ gt.} = 5k^2 \text{ gm.} = 8·4 \text{ gm. for } k^2 = (42/25).$$
$$= \text{ hypothetical mass of the Persian daric.}$$
$$= (1/54) \text{ lb. for lb.} = 453·6 \text{ gm.}$$

I(10) THE DENSITY OF WATER IN TERMS OF THE LIVRE AND THE GREEK CUBIC FOOT

The length of the Greek foot is defined as $(4/10k)$ metre and the water-weight of its cube, therefore, is $(4/k)^3$ kgm. at a density of 1 kgm./litre. In terms of the livre, this water-weight is:

$$(4/k)^3 \times (24/7k^2) = (1536/7k^5) = 60 \text{ livres, for } k^5 = (128/35)$$

As the error in using this value for the fifth power of k is negligible, it appears that the Greek cubic foot of water weighs 60 livres at a density corresponding to 1 kgm./litre, and it would accord with a geodetic hypothesis for the origin of the Greek foot to suppose that its cube in water (rated as a talent of 60 livres) formed the basis for a system of mass.

I(11) THE CUBIC INCH OF GOLD

The density of pure cast gold is commonly published as 19·3 gm./c.c., which is equivalent to $(19·3 \times 252·9) = 4881$ gt./ c. in. and virtually the mean of two close approximations that are more convenient to use in the analysis of ancient metrological evidence. One of these assumes the mass of a cubic inch of gold to be $(9/10)$ tower lb. $= 4860$ gt.; the other rates it $(7/10)$ lb. $= 4900$ gt. These masses I call gold-mina and gold-pound respectively and for convenience I use Gm. and Gp. as their symbols.

$$\text{Gm.} = \text{Gold-mina} = (9/10) \text{ tower lb.} = 4860 \text{ gt.}$$
$$= (25/36) \text{ lb.} = 315 \text{ gm.}$$
$$\text{Gp.} = \text{Gold-pound} = (7/10) \text{ lb.} = 4900 \text{ gt.}$$
$$(\text{Gm./Gp.}) \text{ mass ratio} = (125/126)$$

This is the derivation of the $(25/36)$lb. rating for Gm.: $(9/10)$ tower lb. $= (9/10k)$ lb. $= (25/36)$ lb., for $k = 1.296$. The 315 gm. rating is derived from $(9/10)$ 5400 $(k/20) = 315$, for $k = (35/27)$.

There is evidence implying that a mass equal to that of $(1/20)$ c. in. of gold was used as a rating unit in some Egyptian weights, and the apparent importance of a mass equivalent to 32 such units ($=$ mass of 1·6 c. in. of gold) in the evidence relating to Babylonian weights is such that I call it mina J. Thus J $= 1.6$ Gm. or 1·6 Gp., but I associate it more particularly with the former through the ratings $(10/9)$ lb. $= 504$ gm.

$$\text{Mina J} = 1.6 \text{ Gm.} = (10/9) \text{ lb.} = 504 \text{ gm.}$$
$$= 60 \times (1/54) \text{ lb.} = 60 \text{ darics.}$$

Expressed in grains, the range is from 1·6 Gm. $= 1.6 \times 4860 = 7776$ gt. to 1·6 Gp. $= 1.6 \times 4900 = 7840$ gt. and the approximate mean is 7800 gt. The equivalent of $(10/9)$ lb. is $777\frac{7}{8}$ gt. References to mina J appear in the analyses of the Chinese and the Babylonian evidence relating to weights, also in the monetary evidence relating to the Persian daric.

Among Egyptian units there is one with a mass of 1400 gt. $= (1/5)$ lb. in the category of deben; it is equivalent to the mass of a Sumerian cubic shusi of gold having a density equal to 1200 lb/ c. ft. $= 1600$ lb./Sumerian c. ft.; this corresponds to the density represented by the gold-mina as the mass rating for a cubic inch of gold. I am not aware, however, of any weights that point to this hypothetical connection.

Evidence in support of the gold-pound is less extensive, but sufficient in combination with that relating to the gold-mina to strengthen the hypothesis that the density of gold was used as a basis for some of the ancient weights. For example: in addition to the Egyptian weight already mentioned, which is marked 60 and has a mass of 3 Gm., there is another similarly marked but with a mass of 3 Gp. Also: an inscribed Egyptian weight that is catalogued as the most beautiful in the Petrie collection at University College, London, has a mass of 4 Gp., and mention must be made of a weight that has survived in English metrology—the clove ($=$ half-stone) of 7 lb., which is exactly

10 Gp. in mass.   If this weight really was derived from the mass of 10 c. in. of gold it would give considerable significance to 10 c. in. as a unit of volume and to $(1/32)$ 10 c. in = 5·12 c.c. as a hypo-

One of a pair of cloves in the set of Edward III averdepois weights at Winchester.   Mass = half-stone = 7 lb. = 10 Gp.   The other weights in this set are illustrated in XII(1).   Winchester City Museums photograph.

thetical interpretation of the largest of a set of ancient measures for gold dust that were found in Egypt and are now in the Petrie collection.   The set includes seven measures, obviously intended to form a dimidiated series, and volume ratings proportional to $(10/32)$ c. in. for the largest are in reasonably close accord with the actual measurements:

Actual c.c.	5·30	2·55	1·30	0·65	0·35	0·19	0·10
Hypothetical	5·12	2·56	1·28	0·64	0·32	0·16	0·08

I(12) The (Gold/Silver) Value Ratio

An important aspect of ancient metrology is associated with the (gold/silver) value ratio. Herodotus in his description of the Persian tribute refers to this as 13 but my analysis shows $10k = 12 \cdot 96$ to be the significant number in this case and also in the Persian coinage of that period. Incidentally, this ratio means that a gram of gold was worth 200 gt. of silver; for example, a gold coin weighing 8·4 gm. was worth 20 silver coins weighing 84 gt. each. Similarly, a tower pound of gold was worth 10 lb. averdepois of silver; or 1 lb. of gold was worth 12 livres of silver and so on. It will be observed from these examples that this seemingly awkward ratio was simple metrologically. In Lydia, when Croesus reigned there, the ratio seems to have been (40/3).

Later numismatic evidence would be outside the range of this book, but it may be remarked that this ratio appears to have been 11·2 in the Mohammadan coinage of the eighth century; 11·25 in India at the time of Akbar, and 11·1 (for pure metals) just before Elizabeth's reign began in England: also, some three hundred years earlier (when Henry III revalued his gold penny) it had been the same. As there are 11·1 gt. of pure silver in 12 gt. of sterling,* a (gold/silver) value ratio of 11·1 for pure metals corresponds to a (gold/sterling) value ratio of 12 in terms of the coinage if the gold coin is pure; or to 11 if it is 22 carat. Also, a grain of gold would be worth a half-pennyweight (tower) of pure silver if the value ratio was 11·25, but its value in terms of the currency would depend on the mass of the sterling penny: in Elizabeth's reign the penny was light enough to make pure gold worth about three-halfpence a grain.

I(13) Seed per Acre

Modern farming practice in respect to the quantity of seed sown per acre sometimes provides a useful guide to the interpretation of

---

* This corresponds to 18 pennyweights of alloy in the pound (240 penny-weights) of bullion: sterling is 925 fine. See XIV(6, 7, 8).

ancient evidence relating to areas and capacities when they are connected by seed. The customary sowing of wheat is 2·5 bushels, and of barley from 2 to 3·5 bushels per acre: a plausible rate for either, therefore, is 1 lb. per sq. pole; for wheat, this gives 2·54 bushels at 4·5 st. (= 63 lb.) per bushel; and for barley, 2·86 bushels at 4 st. (= 56 lb.) per bushel.

Barley at 54 lb. (= 50 livres) per bushel and 3·2 bushels per acre is within the above range; and corresponds to 1 livre/sq. pole, or 100 livres (108 lb.) per Roman jugerum. Wheat, or barley, at 100 livres per Egyptian setat is roughly equivalent to 1 lb./sq. pole.

I(14) Summary of Units

The values summarized in these tables define substantially the same magnitudes as those published in one form or another elsewhere, and this was inevitable because all such records reflect the same physical evidence: hitherto, however, there has been no rational basis for preferring one published version to another; the fundamental contribution that my analysis offers in this field is a formula correlating the present and the past. For example, the royal cubit can be visualized as radius in a circle whose circumference, when rated sexagesimally, measures $100k = 129·6$ in.; and this defines the cubit as $(50k/\pi) = 20·6265$ in. Although this number is incommensurable, the formula defining it in terms of k and π is precise. Similarly, Griffith's earlier discovery defined it precisely as $\sqrt{2}$ remen $= 20\sqrt{2}$ digits; but this left the length of the digit uncertain within the narrow range of values derived from measured physical magnitudes.

It was at this point in my analysis that the geodetic aspect of the geometry played a part by providing a very simple definition of the digit, as $(1/54)$ metre; there then remained only the problem of finding the formula linking the metre to the inch.

In the case of the royal cubit, there is a formula appropriate to each of the units in which it may be desired to define its length; but the derived numerical value depends on the rational ratios used to interpret the incommensurables, and also on whether 1·296 or $(35/27)$ is used for k. One set of these values gives the very English-looking length of 20 in. and (5/8)ths of an inch: this is the value in the summary, but others (differing only in the third decimal place) are discussed elsewhere.

Hitherto it has been usual to assume, either that there was no certain standard in ancient metrology or that different magnitudes were the result of indifferent workmanship. My analysis shows that the craftmanship applied to standards of importance was of a very high order, particularly in the case of those weights that have survived in a remarkably good state of preservation on account of the durability of the stone used for them. Also my analysis leads to the hypothesis that there was a system that interlocked the metrology of the civilized world in just the way that the convenience of traders and bankers would demand. It is this commercially interlocking pattern that makes the picture as a whole significant and its parts worth notice.

Several ancient monuments lead to inferences relating to the use of metrological ratios in architectural design, and here also the evidence of convenient approximations is apparent. A case in point is the swimming-bath at Mohenjo-daro: its proportions have a geometrical aspect suggesting 0·76 in. as a significant value for the Greek finger in this connection. Multiplied by 16, this gives a foot of 12·16 in., but the geodetic length in the summary is 12·15 in.; and although this difference in magnitude is negligible, the difference in geometric significance is considerable: it justifies the hypothesis that 0·76 in. and 12·15 in. may be regarded as independently correct interpretations of the Greek finger and the Greek foot respectively.

The values summarized in these tables are derived from the application of the geodetic and geometric hypotheses to evidence that is subsequently set out in sufficient detail to show the very close approximation of the hypothetical to the published measurements. These latter, with the exception of those relating to Gudea's rule, were all made by independent observers.

SUMMARY OF LINEAR UNITS

(Fractions omitted from mm. equivalents)

Inches	*mm.*	
7·07	179	Chinese ch'ih = $(5\sqrt{2})$ in.
9·0	228	Chinese ch'ih = 10 × (9/10) in.
10·8	274	U = 12 × (9/10) in.
10·935	277	(3/4) remen = 12 digits.
11·664	296	Roman ft. = 16 digits.
11·88	301	9 Indus inches.
12·15	308	Greek ft. = (5/6) remen = $(4/10k)$ metre.
12·53	318	$(A/2) = (1/\pi)$ metre. Chinese ch'ih.
12·6	320	1·08 Roman ft.
12·96	329	Assyrian ft. = (8/9) remen = $10k$ in.
13·2	335	Sumerian ft. = 20 × 0·66 in. = 1·1 ft.
13·75	349	Royal ft. = (2/3) royal cubit.
14·14	359	Chinese ch'ih = $(10\sqrt{2})$ in.
14·58	370	Remen = 20 digits.
17·496	444	Roman cubit = 24 digits = (4/9) metre.
17·68	449	(6/7) royal cubit = $(\pi/7)$ metre.
17·71	450	(1/2) $(A\sqrt{2})$ = cubit of the Nilometer.
17·84	453	$\sqrt{(3/2)}$ remen.
18·0	457	20 × (9/10) in.
18·225	463	Greek cubit = 25 digits = (5/4) remen.
19·44	494	Assyrian cubit = (4/3) remen = $15k$ in.
19·8	502	Sumerian cubit = (24/25) royal = (1/10) pole.
20·412	519	28 digits; a version of the royal cubit.
20·52	521	27 Greek fingers. (Finger = 0·76 in.).
20·625	524	Royal cubit = $\sqrt{2}$ remen = $(\pi/6)$ metre.
21·6	548	UU = 2U.
21·87	555	Talmudist cubit = (3/2) remen.
25·06	636	Cubit A = $(2/\pi)$ metre.
25·2	640	2·16 Roman ft. = $(2/\pi)$m.; for π = (25/8).
25·25	641	Palestinian cubit = $\sqrt{3}$ remen.
33·0	838	Akbar's yard = 25 Indus inches = 50 × 0·66 in.

In the Gobi desert Stein found a measure of 9 Indus inches, decimally divided.

At Babylon Oppert found evidence of the use of a linear standard that he identified as U and that I interpret as 10·8 in.: for convenience, I use UU for its double.

Hypothetically, the sacred cubit originated as cubit A. Other versions of it measured 2·16 Roman ft. and $\sqrt{3}$ remen. A square half the area of A^2 has a side $= (1/2)(A\sqrt{2})$.

Royal cubit: alternative ratings, see VI(2):

$20\sqrt{2} = 9\pi$ digits $= (\pi/6)$ metre $= 523\cdot6$ mm.

$20\sqrt{2} = 28\cdot28$ digits $= 20\cdot62$ in.

$20 + (5/8)$ in. $= 20\cdot625$ in. $50(k/\pi) = 20\cdot6265$ in.

Digit $= (1/54)$ metre $= (9/16)k = 0\cdot729$ in.

Shusi.   Assyrian $= (k/2) = 0\cdot648$ in.   Sumerian $= 0\cdot66$ in.

Indus inch $= 2$ Sumerian shusi $= 1\cdot32$ in.
Indus inch $\times$ Greek finger $= 1$ sq. in. See II(6).

Geodetic mile. See I(5).
   $=$ sexagesimal minute of arc on any great circle of the spherical Earth rated 10 million metres in quadrant arc.
   $= (100000/54)$ metres $= 100000$ digits.
   $= 10$ Greek stades $= 6000$ Greek ft. $= 5000$ remens.

English mile $= 320$ poles $= 3200$ Sumerian cubits.
   ,,   furlong $= 40$,, $= 600$,,   ft.

SUMMARY OF LAND UNITS

Roman jugerum $= 1$ myriad Sumerian sq. cubits
Egyptian setat $=$,,   Royal sq. cubits
Palestinian field $=$,,   Palestinian sq. cubits
Hindu nivartana $=$,,   sq. yards
Geodetic acre $=$,,   sq. cubits A

Indian biga $=$ Jugerum $= 100$ English sq. poles. See II(5).
Palestinian jugon $= 13$ Jugera (virtually).   See VII(1).
Babylonian bur $= 25$ Jugera.   See V(10).
English acre $=$ Geodetic acre (virtually).

SUMMARY OF WEIGHTS *
(Omitting fractions)

Grains   Grams

1400 =   90 = (1/5) lb.   An Egyptian deben.
4200 = 272 = 10 unciae.
4860 = 315 = gold-mina = (9/10) lb. tower = (25/36) lb.
4900 = 317 = gold-pound = (7/10) lb.
5040 = 326 = 12 unciae = Roman libra = 14 ores.
5250 = 340 = 12 oz. = (1/2) mina D.
5400 = 350 = tower lb. = (15/16) lb. troy = 15 ores.
5670 = 367 = (3/4) livre = Livre de Charlemagne.
5760 = 373 = troy lb. = (8/7) libra = 16 ores.
6300 = 408 = 15 unciae = (9/10) lb. = Russian funt.
6562 = 425 = 15 oz. = Attic mina.
6615 = 428 = (7/8) livre.
6720 = 435 = 16 unciae.
6750 = 437 = 15 oz. tower.
7000 = 453·6 = 16 oz. = lb. = k lb. tower for k = (35/27).
7200 = 465 = 16 oz. tower = 15 oz. troy = M = 20 ores.
7466 = 484 = (16/15) lb. = k lb. troy for k = (35/27).
7560 = 490 = 18 unciae = 1·08 lb. = Livre = 21 ores.
7680 = 497 = 16 oz. troy.
7776 = 504 = mina J = 60 darics = (16/10) Gm. = (10/9) lb.
7875 = 510 = 18 oz. = (9/8) lb. = Russian bezmen.
8400 = 544 = 20 unciae = water-weight of the sextarius.
10080 = 652 = 24 unciae = 2 librae.
10500 = 680 = mina D = 24 oz. = 25 unciae = 1·5 lb.
15120 = 980 = mina N = 2 livres = 36 unciae = 2·16 lb.
  M = Mercantile pound = 7200 gt.

Uncia = 420 gt.   Tower oz. = 450 gt.   Ore = 360 gt.
Ounce = 437·5 gt.   Troy oz. = 480 gt.

---

* See   I(11) for notes on mina J.
  ,,   V(6)   for other Babylonian standards.
  ,,   VI(10 and 11) for Egyptian kites.
  ,,   IX(11) for small Minoan weights.

In French metrology the livre was rated 2 marks and divided into 16 onces; the 12-once weight = (3/4) livre was called Livre de Charlemagne.

Babylonian talent = 50 livres = 54 lb. = 60 × (9/10) lb.
Euboic talent = 60 gold-minas = 54 lb. tower.
Euboic mina = tower lb.   See VIII(7).

Knossos talent = 64 lb. = 60 × (16/15) lb.

Aeginetan iron spit = 15 unciae = (9/10) lb. = funt.
      handful of six spits = 5 livres.   See VIII(2).
      silver didrachm = (1/40) livre = 189 gt.

Attic silver didrachm = (3/10) oz. = (5/16) uncia.   See VIII(4)
    mina of 100 drachms = 15 oz.

Persian gold daric = 8·4 gm. = (1/54) lb. = (1/60) J. See VIII(8).
    silver siglos = 84 gt.   = (1/5) uncia.

English stones, fotmal, sacks.   See XII(5).

SUMMARY OF CAPACITIES

Water (unless otherwise stated)

Greek c. ft. = Hebrew bath   = 60 livres
Wirksworth dish   = ,,   ,, (lead ore)
Roman amphora = 8 congii = 80 librae = 70 troy lb.
Barrel (beer)   = 50 ,,   = 500 ,,   = 360 lb. = T. sack.
    = 36 gall.   See XII(10).

Troy pint (wheat at 60 troy lb./c. ft.) = troy lb. = (1/60) c. ft.
Sumerian c. ft. = (4/3) c. ft. = 10 troy gall.
Tun = 252 troy gall. = 36 amphorae = 32 Greek c. ft.

Winchester bushel (wheat at 50 lb./c. ft.) = 4·5 st. = 63 lb.
    = 1·26 c. ft. = Assyrian c. ft.

Roman c. ft. = 26·02 litres = (11/12) English c. ft. approx.
Greek c. ft. = 29·403 ,,   = (9/8) Roman c. ft.   ,,
English c. ft. = 28·317 ,,

Indian Metrology: Ancient and Modern

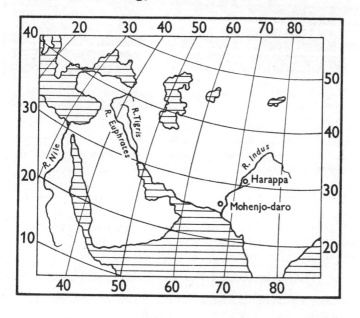

II(I) THE INDUS VALLEY WEIGHTS

INDIAN prehistory revealed an entirely unexpected aspect when the Government of India's archaeological excavations in the Indus Valley disclosed proof of the former existence there of a civilization contemporary and comparable with those of the early periods in Egypt and Mesopotamia.

Among the finds at Mohenjo-daro and at Harappa were many seals with pictographic inscriptions, and numerous stone weights without any rating marks. The former have not yet been deciphered, but the frequency charts that I have constructed from the published masses of the latter show that they were intended to be fractions and multiples of the uncia.

Five of the Indus Valley weights

In the British Museum there are thirteen of these Indus Valley weights, and five of them are shown here; they represent (1/8), (1/4), (1/2), 1, and 2 unciae, and their respective masses are 3·84, 6·6, 13·68, 27, and 54·85 gm.   The lightest and the heaviest are each (4/10) gm. overweight.

The weights that are clearly indigenous are small cubes of chert, and the masses of 288 specimens fall into seven groups with peak frequencies that support these ratings:

(1/16) (1/8) (1/4) (1/2) 1 uncia; also 2 and 5 unciae.

These are the tables from which the charts were drawn.

(1/16) Uncia = 1·7 gm.

1·4	1·5	1·7	1·8	1·9	2·0	2·1	2·2	2·3	gm.
0	0	6	4	2	0	1	0	1	specimens.

(1/8) Uncia = 3·4 gm.

3·2	3·3	3·4	3·5	3·6	3·7	3·8	3·9	4·0	gm.
2	4	17	6	3	0	1	1	2	

(1/4) Uncia = 6·8 gm.

6·6	6·7	6·8	6·9	7·0	7·1	7·2	7·3	7·4	gm.
1	3	22	10	2	0	0	3	0	

(1/2) Uncia = 13·6 gm.

13·4	13·5	13·6	13·7	13·8	13·9	14·0	14·1	14·2 gm.
6	9	23	17	7	7	5	1	1

1 Uncia = 27·2 gm.

26·6	26·8	27·0	27·2	27·4	27·6	27·8	28·0	28·2 gm.
1	5	11	31	17	5	4	2	1

2 Unciae = 54·4 gm.

53·6	53·8	54·0	54·2	54·4	54·6	54·8	55·0	gm.
1	1	6	0	8	4	0	1	

5 Unciae = 136 gm.

134	135	136	137	138	gm.
1	3	6	0	1	

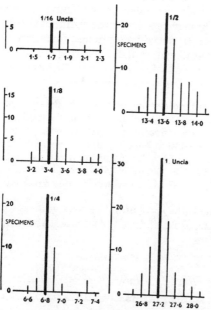

Frequency charts of weights found at Mohenjo-daro. The horizontal scale shows mass in grams. The vertical scale shows number of specimens.

II(2) THE RETTI

Retti is the name by which the seed of the *Abrus precatorius* is known in connection with Indian metrology: this comically pretty spherical product of nature, with its parti-coloured exterior (one half black, the other bright red), has long done service as a weight but its mass depends on its moisture content. Twenty specimens (generously supplied by the University School of Botany at Oxford) weighed 20 gt.; fifteen of them weighed 15 gt.; a random pair weighed 2 gt. but these seeds had been dry for years: other investigators have published masses between 1·5 and 2 gt. depending on the freshness.

The earliest documentary reference to the use of this seed as a weight appears in a Sanskrit work called the *Laws of Manu*; a comprehensive code compiled (*c.* 500 B.C.) by the 'supposed ancestor of the human race.' In the eighth chapter, a long series of mass ratios is introduced by a promise to declare the 'technical names of certain quantities of copper, silver, and gold which are generally used on earth for the purpose of business transactions among men.' Buhler's translation proceeds thus:

> The very small mote which is seen when the sun shines through a lattice, they declare to be the least of all quantities and to be called a transarenu (a floating particle of dust). Know that eight transarenus are equal in bulk to a liksha (egg of a louse); three of these to one grain of black mustard and three of the latter to a white mustard seed. Six grains of white mustard are one middle sized barleycorn and three barleycorns one krishnala.

This krishnala is the retti, and it seems significant that its rating is $1000k = 1296$ specks of dust. The ratios above mentioned can be tabulated thus:

Mustard seed

dust	liksha	black	white	barley	retti
1296 =	162 =	54 =	18	= 3	= 1

A trial weighing of Kenya barley, which is of medium size, gave 4·34 gm. as the mass of 100 barleycorns; this makes $1000(k/3) =$

432 tenths of a milligram a plausible mass rating for the barley-
corn; and as this corresponds numerically to its rating by Manu in
terms of the speck of dust, the mass rating of the dust speck can
be regarded as (1/10) milligram: in terms of this, the proportional
retti is seen to have a mass of (k/10) gm. = 2 gt.

The remainder of Manu's mass ratios are:

retti		masha		suvarna		pala		dharana
3200	=	640	=	40	=	10	=	1

The tradition of a pala rated 320 retti survived for at least 1600
years, for it appears in Bhascara's *Lilavati* published *c.* A.D. 1150.

II(3) THE TOLA, THE SEER, AND THE MAUND

In 1833 James Prinsep (Assay Master at Calcutta) wrote a
letter to the Mint Committee recommending the adoption of the
southern rupee; and suggesting that 180 gt.* should be standardized also as the mass of the tola in a system of weights having a
seer of 80 tolas representing 2·5 lb. troy, and a maund of 40 seers
that would be 100 lb. troy.   His scheme was adopted for government transactions, but did not acquire much wider vogue until its
use by the railways caused these units to becomes generally known
as the railway weights.

The Railway Weights (used in North India)

1 maund	= 40 seers	= 3200 tolas	= 100	lb. troy	
	1 seer =	80 ,,	=	2·5 ,,	,,
		1 tola =	(1/32) ,,	,,	= 180 gt.
			= rupee weight		

In 1834—the year after that in which Prinsep wrote his letter—
the troy standard was lost in the fire that destroyed the Houses of
Parliament; and the averdepois pound, after a considerable
interval of inquiry, became the legal standard in 1878.   In this
interval an attempt to substitute the kilogram (2·205 lb.) for the

---

* At that time the mint at Ferruckabad was coining rupees weighing
180·234 gt.

seer of 2·5 lb. troy = 2·057 lb. was made in India: a clause in the unrepealed (but still-born) Weights and Measures Act of 1870 reads 'the primary standard of weights shall be called a seer and shall be equal to a kilogramme.'

In 1932 the railway weights were adopted as standard by the Bombay Government, but the system previously in use there was:

Bombay scale of mass (prior to 1932)

$$1 \text{ candy} = 20 \text{ maunds} = 800 \text{ seers} = 560 \quad \text{lb.} = 5 \text{ cwt.}$$
$$1 \text{ maund} = 40 \quad ,, \quad = 28 \quad ,, = 1 \text{ qr.}$$
$$1 \text{ seer} = (7/10) \quad ,, = \text{Gp.}$$

This Bombay system was the result of adapting an earlier indigenous scale to the averdepois standard; and, according to Strachey,* the old Bombay seer was reckoned 30 pice and the maund of 40 such seers was nearly 28 lb. In Southern India, also according to Strachey, 'the original unit was the pagoda'; it is important, therefore, to note that if the Madras seer of (3/4) lb. troy is rated 80 pagodas, and the Bombay pice is equated to 3 pagodas, then the Bombay old seer of 30 pice will weigh (9/10) lb. tower and the proportional maund of 40 such seers will be 36 lb. tower = 27·7 lb. This easily could have been changed to 28 lb. = 1 qr. (under British commercial influence) and if this actually happened it would mean that an indigenous seer equal to the gold-mina was changed to equality with the gold-pound.

Bombay (hypothetical) old scale of mass

$$1 \text{ maund} = 40 \text{ seers} = 1200 \text{ pice} = 36 \text{ tower lb.} \quad = 27·7 \text{ lb.}$$
$$1 \text{ seer} = 30 \quad ,, \quad = (9/10) \text{ tower lb.} = \text{Gm.}$$
$$1 \quad ,, \quad = 3 \text{ Madras pagodas}$$

Madras (hypothetical) old scale of mass

$$1 \text{ maund} = 40 \text{ seers} = 3200 \text{ pagodas} = 30 \text{ lb. troy}$$
$$1 \quad ,, \quad = 80 \quad ,, \quad = (3/4) \text{ lb. troy}$$
$$1 \text{ pagoda} = (1/100) \text{ lb. troy} = 54 \text{ gt.}$$

---

* In a pamphlet published in 1857; after an official inquiry into Indian weights and measures by a committee of which he had been chairman. Quoted in the **Silberrad report**.

Madras (modern) scale of mass

1 candy = 20 maunds = 800 seers = 600 lb. troy
  1 maund = 40 ,, = 30 lb. troy = 24·685 lb.

But the maund is frequently reckoned as 25 lb. and the candy as 500 lb.

II(4) THE INDUS INCH

The linear unit that I call the Indus inch is engraved on a broken piece of shell found at Mohenjo-daro (mound of the dead) in the Indus Valley: the find was made in 1931 during archaeological excavations carried out under the aegis of the Government of India. The fragment of shell measures 6·62 by 0·62 cm. and shows nine parallel lines, 0·264 in. apart, cut with a fine saw.

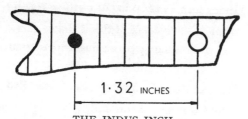

1·32 INCHES

THE INDUS INCH
Rough sketch of the piece of shell
found at Mohenjo-daro

One of these lines is distinguished by a circle; another, five spaces distant from it, is marked by a large dot: it is the length of 1·32 in. between the circle and the dot that I call the Indus inch. These are published measurements,* based on a report by Sir Flinders Petrie; they make the Indus inch exactly equal to 2 Sumerian shusi.

Indus inch = 2 Sumerian shusi

Mohenjo-daro (27° 19′ N.; 68° 8′ E.) is about three and a half miles west of the Indus, but it is probable that this distance was

---

* *Further Excavations at Mohenjo-daro*, by E. J. H. Mackay, vol. i, pp. 348, 404 (1938).

caused by an eastward shift of the river-bed and that the site was made uninhabitable thereby. Aristobulus, on a mission for Alexander the Great, reported that this eastward movement of the Indus had abandoned to the desert more than a thousand towns and villages that formerly had thrived in its valley: of such places Mohenjo-daro must have been one of the more important; the archaeological evidence indicates that it was rebuilt at least seven times in a period of about 500 years.

Evidence of the survival of the Indus inch in India seems apparent in the 25-Indus-inch rating that can be applied to the 33-inch North India gaz; this linear unit was traditional in the time of Akbar, and standardized by the British after careful examination of the evidence.

A formula for the length of the Indus inch (derived from an analysis of the bath at Mohenjo-daro) can be expressed as follows:

$$\text{Indus inch} = (\sqrt[4]{3}) \text{ in.} = 1\cdot316 = 1\cdot32 \text{ in.}$$

The metrological significance of this is that:

$$\text{Indus inch} \times \text{Greek finger} = 1 \text{ sq. in.}$$

II(5) Akbar's Yard, and the Biga

In 1825 the Government of India, having considered the results of an inquiry into the local measurements of the indigenous land-unit called biga, 'adopted an arbitrary value of 33 inches as the Ilahi gaz.' * This Indian yard had been established after a similar inquiry by Akbar, who came to the throne in 1556; but Akbar's 'attempts at unification were not successful because of the decline of the central power and the internal confusion that followed.' †

Akbar's yard = 33 in. = 25 Indus inches
(The Ilahi gaz) = (1/6) English pole

This link with the Indus inch is strong evidence in support of the probability that the decision to standardize 33 in., although arbitrary, accurately perpetuated ancient tradition; but the

---

* *Chronicles of the Pathan Kings of Delhi*, by Edward Thomas.
† Silberrad Committee Report, p. 11.

decisive factor in its favour is that it makes the biga equal in area to the Roman jugerum, and this interpretation is unconsciously supported by this passage in the Committee's report:

> The chief indigenous measure of area is the biga, which is said to have been the area which a pair of oxen can plough in a day.

It is regrettable, perhaps, that this ancient significance was not recognized; because the square pole would have provided a ready-made minor unit: instead, the authorities decided to introduce the English acre subdivided into hundredths called decimals.

The metrological ratios in the report are:

Jarib = 20 lath = 60 gaz = 1440 tasu = 5760 pan
Biga = sq. jarib = 60² sq. gaz

In terms of the Ilahi gaz measuring 33 in., the length of the jarib is 10 English poles; and the area of the biga is 100 sq. poles = Roman jugerum. This gaz and this biga survive in North India but not in other districts.

North India

Gaz = 144 jow = 33 in. = 25 Indus inches = (1/6) pole.
Kos = 4000 gaz = 2 + (1/12) miles.
Biga = 400 baus = 3600 sq. gaz = 100 sq. poles = jugerum.

Bombay

Gaz = 1·5 hath = 27 tasu = 27 in.   Hath = 18 in.
Biga = 20 pand = 400 kati = 16000 sq. hath = 4000 sq. yds.

Madras

Cubit = 18 in.   Span = 8 in.   Kos = 8000 cubits.
Kani = 100 guli = 160² sq. cubits = 6400 sq. yds.

II(6) THE GREAT BATH AT MOHENJO-DARO

The principal architectural discovery, revealed during the Government of India's archaeological excavations at Mohenjo-daro, was the Great Bath; which is thus described by Sir John Marshall in *Mohenjo-daro and the Indus Civilization* (p. 24).

> The Great Bath . . . was part of what appears to have been a vast hydropathic establishment and the most imposing of all the

remains unearthed at Mohenjo-daro. Its plan is simple; in the centre an open quadrangle with verandahs. . . . In the midst of the open quadrangle is a large swimming bath, some 39 ft. long by 23 ft. broad and sunk about 8 ft. below the paving of the court, with a flight of steps at each end, and at the foot of each a low platform . . .

In *Further Excavations at Mohenjo-daro* (p. 131), E. J. H. Mackay gives these details:

It is constructed of specially cut bricks of varying sizes: the dimensions of the bath are:

Inches		*Inches*	
West side	= 472	South end	= 275
East ,,	= 471	North ,,	= 280·5

This is relatively accurate lay-out for brickwork and the slight discrepancy is amply atoned for by the careful finish of the masonry—a finish so good that the writer has not seen its equal in any ancient work.

It seems certain that the intended area of the bath was 100 sq. yds. = 129600 sq. in.: actual measurements give:

Mean length × south end = 129662·5 sq. in.

If this interpretation is correct, then it is also probable that it represented (1/100) nivartana in terms of an area so called in the *Lilavati*.

The proportions of the bath suggest that the intended length of the diagonal was twice the breadth: if so, then:

(length/breadth) = $\sqrt{3}$

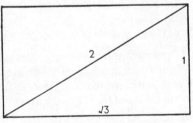

Geometric plan of the bath.

Also, if the intended area was 100 sq. yds. = 360^2 sq. in.; then the length might have been $360x$ and the breadth $360y$ where:

$$x = (\sqrt[4]{3}) \text{ in. and } y = (1/x) \quad \therefore \; xy = 1 \text{ sq. in. and } (x/y) = \sqrt{3}$$

This arbitrary hypothesis, of equal numerical ratings for length and breadth (but in terms of reciprocal units x and y), seems to be a fruitful guess; for the units so disclosed are:

$$x = 1\cdot316 = 1\cdot32 \text{ in.} = \text{Indus inch}$$
$$y = (1/1\cdot316) = 0\cdot76 \text{ in.} = \text{Greek finger}$$

It appears, therefore, that the length and breadth of the bath can be rated:

Length = 360 Indus in.   Breadth = 360 Greek fingers

This unexpected result lends strong support to the hypothetical importance of the English inch in the analysis of ancient metrology, and it suggests $(\sqrt[4]{3})$ in. as a plausible origin for the Indus inch; with 1·32 in. as a simplified version adapted to other scales. Also, it shows 0·76 in. as an appropriate independent rating for the Greek finger.

II(7) THE HINDU NIVARTANA

In the metrology of Bhascara's *Lilavati* * (c. A.D. 1150) there is a land-unit called nivartana that measures 20 × 20 bamboo poles, with a pole measuring 10 cubits.

In the Silberrad Committee's report there is a cubit of 18 in.: and there is the strong evidence of the ancient use of a 9-in. ch'ih in China.   In terms of an 18-in. cubit, the area of the nivartana is 1 myriad sq. yds. = 60^4 sq. in.   Support for this rating also appears in the area of the bath at Mohenjo-daro, and in the geometric series of acres in Britain.

---

* See XV(1): Colebrooke.

III

Geometric Aspects of Chinese Metrology

III(1) THE CH'IH

ACCORDING to Wang Kuo-wei,* a ch'ih of 9 in. was in use during the Tchou Dynasty (*c.* 1100-256 B.C.), and

a standard of 13·5 = (3/2) × 9 in. was still in use during the Ming Dynasty (A.D. 1368–1644). An inscription on a bronze hu (bushel), made during the reign of the usurper Wang Mang (A.D. 9–23), states that its depth should be 1 ch'ih; and its actual measurement is 9 in. The history of the Han Dynasty (206 B.C.–A.D. 220) refers to the pu coin (first cast in A.D. 14 and now nicknamed trousers cash) as (1/4) ch'ih in length; and extant early specimens measure (1/4) × 9 = 2·25 in. From the ruins of the watch-towers used by soldiers guarding the caravan route across the desert, Stein † recovered documents dated in the first century A.D. and also two 9-in. ch'ih measures decimally divided into t'sun of (9/10) in.

Pu money ‡

Some pieces of silk were also among Stein's finds: one of these has its width marked on it in a manner that means 2 ch'ih 2 t'sun;

* 'Chinese foot-measures of the past nineteen centuries,' by Wang Kuo-wei. Translated from the Chinese by A. W. Hummel. *Journal of the Royal Asiatic Society* (North China Branch), vol. lix (1928), p. 111.

† *Serindia. A detailed report of explorations in Central Asia and Western-most China, carried out and described under the orders of H.M. Indian Government,* by Sir Aurel Stein (1921).

‡ Heberden Coin Room, Ashmolean Museum.

and if this ch'ih is interpreted as 9 in. decimally divided, then the intended width was:

2 ch'ih 2 t'sun = 22 t'sun = 19·8 in.

= Sumerian cubit.

The published width is 50 cm. = 19·685 in.

Referring to the payment of taxes in silk, Wang Kuo-wei states that 'the standard of value was the p'i (bolt) which was required to measure 40 ch'ih in length and 2 ch'ih 2 t'sun in width.'

It is quite certain that 9 in. was not the only ch'ih in ancient China: Terrien de Lacouperie * says that 'according to the regulations of Tchou li, the bronze knives must be cast 1 ch'ih in length'; and he gives the following statistical evidence relating to this early form of Chinese money.

Average length in inches	7·4	7·1	6·8
Specimens in group	20	81	25

His comment is 'we must look upon the 81 cases of 7·1 in. as representing the standard.' It seems certain that this was the ch'ih in which it was customary to express a man's height. Confucius (551–479 B.C.) was 9·6 ch'ih and regarded his father's height of 10 ch'ih as the culmination of tallness. Mencius (372–290 B.C.) also considered it of sufficient interest to remark that one of his disciples grew to a height of 9·4 ch'ih.

It was the double of this standard that survived into modern times as a ch'ih of 0·36 metre: evidence in support of this ch'ih and of its multiple the itinerary li of 1800 ch'ih is to be seen in a

Knife money †

---

* British Museum Catalogue of Chinese Coins.
† Heberden Coin Room, Ashmolean Museum.

little book written by Bodley's librarian, Thomas Hyde, in the reign of James II; it was published in 1688 under the title *Epistola de mensuris et ponderibus Serum seu Sinensium* and is based on information supplied by a Chinese undergraduate at Oxford University. In particular, it refers to an itinerary distance of 10 li (called p'u) as being 'a little more than 4 English miles': in terms of a ch'ih of 0·36 metre and a li of 1800 such ch'ih the p'u would measure 4·05 miles in terms of a mile reckoned as (5/8) kilometre. Hyde's statement reads:

Maxima longitudinis aut distantiarum apud Chinenses mensura est can, quae facit 8 p'u seu 80 li; quod apud eos est ordinarium iter diurnum, quantum Sinenses ambulando conficiunt. Adeo ut unum p'u faciat 10 li, id est 4 milliaria Anglica et paulo plus.

∴ Can = 8 p'u = 32·4 miles = 'a day's journey.'

The ancient Chinese linear scale was:

$$Hao \quad Li \quad Fen \quad T'sun \quad Ch'ih \quad Chang \quad Ying$$
$$10^6 = 10^5 = 10^4 = 1000 = 100 = 10 = 1$$
$$\text{Itinerary li} = 18 \text{ ying} = 1800 \text{ ch'ih}$$

The traditional decimal aspect of this scale expressed the national regard for 10 as the perfect number, but the itinerary li introduced a sexagesimal element.

The ch'ih of 0·36 metre was called kung (official) ch'ih in order to distinguish it from the shih ch'ih of 0·32 metre that was customary in the markets. While these metric values do not necessarily agree exactly with the original lengths they were intended to perpetuate, nevertheless their (8/9) length ratio may reflect the ancient rule for squaring the circle by squaring eight-ninths of its diameter; if it does so then the square ch'ih was intended to represent the area of a circle 1 kung ch'ih in diameter.

Now if the kung ch'ih of 0·36 metre is interpreted as 0·359 metre it can be represented as $10\sqrt{2} = 14·14$ in., and the length of the knife money becomes $5\sqrt{2} = 7·07$ in. instead of 7·1 in. The evidence of the knife money, therefore, supports this interpretation of the kung ch'ih. A circle of this kung ch'ih radius would have an area of $(10\sqrt{2})^2\pi = 200\pi$ sq. in. = square on cubit A, which gives $(A/2) = (1/\pi)$ metre as the interpretation to be applied to the shih ch'ih of 0·32 metre on the hypothesis that the square on the shih ch'ih was intended to have an area equal to that of a circle of kung ch'ih diameter.

This interpretation of the shih ch'ih implies that it had a geodetic origin; also that a circle of 1 kung ying = 100 kung ch'ih = 1000√2 in. radius would have an area equal to that of the square on 100A, that is to the geodetic acre.

$$\text{Kung ch'ih} = 0\cdot36\,\text{m.} = 0\cdot359\,\text{m.} = 10\sqrt{2} = 14\cdot14\,\text{in}$$
$$\text{Shih ch'ih} = 0\cdot32\,\text{m.} = (1/\pi)\,\text{m.} = (A/2) = 12\cdot53\,\text{in.}$$
$$0\cdot32\,\text{m.} = (1/\pi)\,\text{m., for}\ \pi = (25/8)$$

The probable validity of this geometric interpretation is enhanced by the well-established presence of the 9-in. ch'ih in the historical evidence relating to ancient Chinese metrology: superficially, this 9-in. standard may seem to confuse the issue, but in fact it enlarges the geometric aspect by representing the diameter of the circle whose quadrant-arc is equal to the ch'ih of 5√2 in. The significance of this relationship is that the area of any circle can be represented by that of a rectangle having sides respectively equal to its diameter and its quadrant-arc. The geometric equation in this case is based on the hypothesis in I(3) expressed by:

$$\text{Quadrant (chord/arc) length ratio} = (2\sqrt{2})/\pi = (9/10)$$
$$\therefore\ (5\sqrt{2})/9 = (\pi/4)$$

From the Tchou Dynasty to the Ming Dynasty is an interval of more than 1500 years, but during the whole of this period the 9-in. standard seems to have maintained its influence in Chinese metrology. During the Ming Dynasty there was a ch'ih of exactly $(3/2)9 = 13\cdot5$ in. and during the intervening dynasties a measure of 11·25 in. appeared, which draws attention to this proportion:

$$9\ :\ 11\cdot25\ :\ 13\cdot5\ ::\ 4\ :\ 5\ :\ 6$$

The common factor in this series is 2·25 in., the length of the pu coins.

Two examples of the 11·25-in. standard were included in a set of six ivory measures given (as a votive offering, by a dowager empress of Japan in A.D. 757) to the Tung Ta monastery. They are distinguished in pairs by the colours white, red, and green corresponding to three different lengths that evidently were intended to form a geometric progression; that is to say the rectangle white × green was intended to be equal in area to the square on the red.   On this hypothesis the intended lengths were 11·25, 11·6 (nearly), and 11·92 in.: the published lengths are:

White = 11·25   Red = 11·646   Green = 11·917 in.
  ,,   = 11·25   ,, = 11·606   ,,   = 11·917 in.

Possibly some of these lengths have further implied significance: for example, the square on the green is close to one-third the area of the square on the royal cubit.

Wang Kuo-wei associated these six examples with the Tang period, and he mentions two more: one of these is a beautifully carved specimen of the kind that had to be presented to the emperor on the second day of the second month each year. Documentary evidence relating to this ceremony appears in the 'Shan Shu Ling' section of the *Ta T'ang Liu Tien*.   In no other country has metrology been the focal point of a court function.

About 1929 the Government in China adopted the metre as a new kung ch'ih and changed the shih ch'ih to (1/3) metre.   The linear scale established by this change is:

	li	*fen*	*ch'ih*
Kung	1 mm.	1 cm.	1 metre
Shih	(1/3) mm.	(1/3) cm.	(1/3) metre

Itinerary kung li =   1 kilometre
    ,,   shih li = (1/2)   ,,   = 1500 new shih ch'ih.

III(2) CHINESE WEIGHTS

New system: Kung chin = kilogram
            Shih chin = (1/2) kilogram = 500 gm.

Formerly: Shih chin = (4/3) lb.   = 604·8 gm. = Catty
            Shih liang = (4/3) oz.   = (3/40)J. = Tael

Probably the catty acquired its mass rating as the result of adjusting an earlier standard to English commercial usage, but I have not been able to find such a standard in the confusing evidence of the coinage. Lacouperie, however, gives the following names and mass ratios based on his study of the documentary evidence and he suggests 780 gt. as an appropriate value for the yuen. This is equivalent to suggesting mina J as a rating for the hwan and makes the hypothetical mass of the lighter kin equal to (1/25) Gm., which is the mass of one of the Egyptian gold-sign kites and also the mass of one of the smaller Minoan weights. That there was some form of gold basis in the ancient Chinese system of mass seems certain from Lacouperie's reference to the kin as:

'The gold unit of the Tchou financial laws, being equal to a cubic inch of the precious metal.'

Tchou	Hwa	Tche	Kin	Yuen	Kin	Lut	Hwan	
1920 =	320 =	160 =	40 =	10 =	4 =	2 =	1 =	J
	Hypothetical		(1/25)		(2/5)			Gm.
				6	15	30	60	Darics

IV

Links with the Past in Russian Metrology

PETER THE GREAT is said * to have reduced the Russian linear scale by about one per cent in order to make the sajen of 6 fuss exactly equal to 7 English feet. If 7·07 ft. = 5√2 ft. was its original length, then the original fuss can be rated 14·14 = 10√2 in. and thereby associated with the Chinese old kung ch'ih; similarly, 6 versts (= 18000 fuss) would then be equal to the Chinese p'u of 10 li rated 18000 ch'ih. On the same basis, the original square sajen measured 50 sq. ft.

Versock	Fuss	Sajen	Verst	Dessatine Sq. Sajen	
24000	= 3000	= 500	= 1	Crown	= 2400
Also; Arshin = 2 fuss				Common	= 3200

The published mass of the Russian pood is 36·11 lb., which suggests that it originated as 36 lb. = 50 librae. This would make the zolotnik of (3/20) oz. equal in mass to the Attic drachm introduced by Solon. In Greece, the mina of 100 such drachms weighed 15 oz.; in Russia, the funt of 96 zolotniks weighed 15 unciae; and this, hypothetically, was the mass of the drachma of Aeginetan iron spits. The name zolotnik is derived from a root meaning of gold.

Zolotnik	Funt	Pood		
3840	= 40	= 1	Pood = 36·114 lb. (Published)
	96	= 1	= 36 ,, (Hypothetical)
			= 50 librae ,,
Loth = 3 zolotniks			∴ Bezmen = (9/8) lb. = 18 oz.
Doli = (1/96) ,,			Funt = (9/10) ,, = 15 unciae
Bezmen = (5/4) funt			Zolotnik = (3/20) oz. = (5/32) ,,
	Garnez = 0·7218 gall. (Published)		
	= 0·72 ,, = Roman congius		

---

* By N. T. Belaiew in *Seminarium Kondakovianum*.

V

Babylonian Metrological Evidence

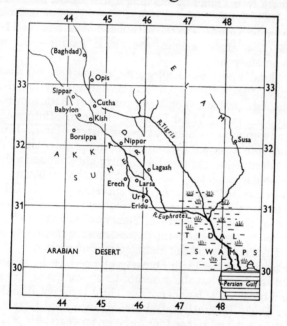

V(1) INTRODUCTION

IN THE cuneiform inscriptions, Sumer and Akkad are the names of Babylonia; * and the Sumerians dominated all that region until the rise of Semitic power in Akkad.   Archaeological excavations

---

* Babylon itself was a relatively unimportant place until it acquired renown through splendour as the capital city in the reign of Nebuchad-nezzar II (605–562 B.C.).

at the sites of some of the Sumerian cities in southern Mesopotamia have revealed evidence relating to incidents in the lives
of their inhabitants; for example, a boundary canal was the
frequent cause of strife between the peoples of Lagash and Umma,
and the victory of Eannatum of Lagash is recorded on a monument called the Stele of the Vultures.   Entemena, fifth governor
of Lagash in this period, also was victorious over Umma but he
added more to his fame by building a new canal: his name and
that of High Priest Dudu are on a silver vase of metrological
import that he dedicated to the patron god of his city; and Dudu's
name is on a stone weight—the oldest extant—now in the
Ashmolean Museum.

   These local rulers owed allegiance to the king of Kish, but
Urukagina of Lagash revolted: he was defeated by Lugalzaggesi
of Umma, who embarked on a campaign that made him the first
and only king in the Third Dynasty of Erech.   Both men were
taking advantage of trouble at headquarters; for Ur-Ilbaba of
Kish had been attacked by Sargon, a Semitic chieftain in Akkad
who was destined to conquer the country and to found a dynasty
that endured for a century before being overthrown by an invasion
from Gutium.   It was during this supremacy of the Guti that
Gudea (c. 2175 B.C.) became the tributary ruler of Lagash: 'he
was one of the great figures in Sumerian history' and is commemorated by eighteen extant statues, most of them decapitated;
two of these happen to be of metrological importance.

   Utu-khegal, of the Fifth Dynasty of Erech, led a revolt that
expelled the Guti; but the governor he appointed at Ur expelled
him, and became the founder of the Third Dynasty of that name.
His son Shulgi (formerly called Dungi) succeeded to the throne
about 2100 B.C. and reigned for forty-seven years: his period was
one of great prosperity.   Probably it included some administra-
tive act in relation to metrology, for his name is mentioned on a
weight (mina N) that was made at least 1500 years later.

V(2) GUDEA'S RULE

In 1881, when excavating the ruins of Lagash, de Sarzec found eight headless statues of Gudea, governor of that city about 2175 B.C. These statues are now in the Louvre and two of them show Gudea seated, with a tablet on his lap.* One of these tablets is blank, except for a graduated rule near the edge and a stylus ready for right-handed use: on the other tablet, these instruments

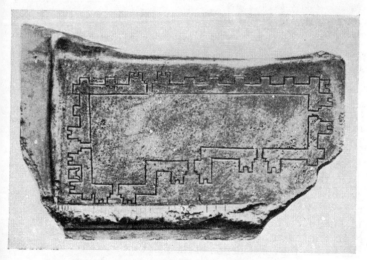

Gudea's tablet showing the ground plan of a temple. The graduated rule near the front edge is better preserved on another tablet that is otherwise blank.

are accompanied by a well-executed architectural plan. It has been suggested that these statues were intended to portray Gudea as a supplicant for inspiration, and dedicating his achievement.

The scale on the rule has 16 nominally equal divisions, with a total length of 269 mm.; the average length of a division, therefore, is 16·8 mm. = 0·66 in.

In the Louvre, the statues of Gudea are identified by the letters

* See page 4.

A to H: those with the tablet (F and B) are called 'L'Architecte à la règle' and 'L'Architecte au plan.' There is a copy of B in the British Museum (No. 91025) but it is on F that the rule is better preserved. In *Découvertes en Chaldée* (1884–1912) de Sarzec and Heuzey included the full-scale photograph from which I made these measurements * of the 16 divisions on the rule:

Portion of Gudea's rule, natural size

17·5   17·7   16·5   16·8   16·0   17·7   16·6   17·2   mm.
16·5   16·5   16·5   17·3   16·2   17·0   16·5   16·5   mm.
Total length = 269 mm.   Average division = 16·8 mm. = 0·66 in.
= Sumerian shusi.

On the evidence of the cuneiform texts, the cubit contained 30 shusi: if, therefore, the length of the division on Gudea's rule is called a Sumerian shusi, the proportional Sumerian cubit was 19·8 in., or (1/10) English pole. The proportional Sumerian foot of 20 shusi measures 13·2 in.; and there is evidence of the use of this foot in the construction of the first century (A.D.) fort known as the Caburn,† on the Sussex downs near Lewes.

---

* In order to confirm the accuracy of these measurements, I sent a full-scale copy to the Louvre with a request that it be compared with the original: the reply from the Conservateur en Chef des Antiquités Orientales reads:
    Je vous renvoie, avec les vérifications demandées par votre lettre, la règle qui est reproduite sur la statue de Gudea. J'ai indiqué deux coches supplémentaires invisibles sur le photo, entre les deux cassures. Le reste est exact.
These two additional notches mark one-tenth of a division, and do not in any way affect my interpretation of the length of the division itself.

† *Geographical By-ways*, by Sir Charles Arden-Close (1947). The holes were found in 1877 by Pitt-Rivers but it was not until 1938 that their distance apart was measured. Also see XV(1) Petrie; (4) McCaw.

V(3) THE ASSYRIAN FOOT

Measurements, made in 1853 by Oppert at the site of the ruins of Babylon, revealed evidence of the use of a linear unit that he published as 329 mm. and called the Assyrian foot. This length can be interpreted as $10k = 12\cdot96$ in., or as $20 \times (8/9)$ digits; which makes the length of the Assyrian shusi $= (8/9)$ digit $= (k/2) = 0\cdot648$ in.

Evidence of the use of the proportional cubit of 30 such shusi is to be seen in the metrological ratios of the cuneiform texts; there an area called bur measures 25 Roman jugera in terms of this cubit.

It is also interesting to note that the cube of the Assyrian foot defined as $(10k/12)^3$ c. ft. measures $1\cdot26$ c. ft., for $k^2 = 1\cdot68$; and this can be regarded as a prototype Winchester bushel.

V(4) MINA D

The oldest extant weight is in the Ashmolean Museum * at Oxford, and I call it mina D because its inscription—one mina of wages in wool—is signed by Dudu the high priest at Lagash *c.* 2400 B.C. Its mass was published by Langdon † as '680·485 gm. or about a pound and a half,' but as the excess is less than one part in 8000 this weight can be regarded as an expression of the following significant rating:

$$1\cdot5 \text{ lb.} = 24 \text{ oz.} = 25 \text{ unciae.}$$

Evidence of the ability of certain craftsmen to make accurate metrological standards (for important people) is one of the most notable new aspects of remote antiquity, and this particular instance of it is especially interesting because of the mass ratio of mina D to the English pound.

---

* Exhibit No. 1921.870.
† In the *Journal of the Royal Asiatic Society*, 1921, p. 575.

MINA D

The oldest extant weight, *c.* 2400 B.C. It is inscribed with the name of
Dudu, high priest at Lagash in the reign of Entemena.   Mass = 1·5 lb. =
24 oz. = 25 unciae.   The height is exactly 4 in.   Ashmolean Museum
photo of exhibit 1921.870.

MINA N

The inscription refers to Nebuchadnezzar II (605–562 B.C.); also to Shulgi of the Third Dynasty of Ur, *c.* 2100 B.C. Mass = 2 livres = 36 unciae = 2·16 lb. Height = 3 + (11/16) in. British Museum photo of exhibit 91005.

V(5) Mina N

In the British Museum there is a weight * that I call mina N because its cuneiform inscription certifies it to be a copy of a weight that Nebuchadnezzar II (605–562 B.C.) made according to a weight that belonged to Shulgi (c. 2100 B.C.) of the Third Dynasty of Ur. This inscription, therefore, is proof of intention to preserve a standard of mass for at least 1500 years, but it has not previously been noticed that this standard was the double-livre.

Mina N has been weighed twice. Its mass was published by Airy, as 15100 gt.; and by Belaiew, as 978·3 gm. The mass of 2 livres (based on the mass of the livre established when the metric system was introduced) is 979·38 gm.; but in this analysis its metric rating is $(7/24)k^2$ kgm. = 980 gm., for $k^2 = 1·68$. In terms of the pound, its rating is 1·08; which makes the mass of mina N = 2·16 lb. Or, it can be regarded as equal to 60 shekels in terms of a shekel = 0·036 lb.; and this happens to be the mass of a cubic inch of water at about 80° F.

Among the relationships that support this interpretation, are these:

Mina N = 2 French livres = 3 Roman librae.

Libra	Tower lb.	Troy lb.	Livre	
5040	5400	5760	7560	gt.
14	15	16	21	ores

Libra = 12 unciae   ∴ Livre = 18 unciae = 1·08 lb.

= 18 × (24/25) oz.

∴ (Uncia/Oz.) mass ratio = (24/25)

V(6) Some other Babylonian Weights

Next in antiquity to mina N is a weight inscribed:

Fifteen shekels of the god Ningírsu,
made by Urukagina, king of Girsu.

---

* No. 91005.

This rating is equivalent to (1/4) mina; and as the published mass is 119·3 gm. my interpretation of the proportional mina is 476·28 gm. = 1·5 Gp.: the mass of the proportional shekel is (1/40) Gp. and that of the weight itself is (3/8) Gp.

, Two other weights are virtually in this scale, but are more appropriately rated in terms of the gold-mina of 315 gm. than in that of the gold-pound of 317·5 gm. Their masses are 236 gm. = (3/4) Gm., and 946 gm. = 3 Gm.; but this latter weight has 2 as its rating mark and so its mina is 1·5 Gm. It is inscribed:

> Two minas. Tiglath-pileser, king of the country.

¡This evidence for 1·5 Gm. as the mina in this weight supports 1·5 Gp. as the mina in Urukagina's weight: metrologically, these minas are equal in significance, for both imply a mass equal to that of 1·5 c. in. of gold.

Three of the marked weights published with fractional ratings can be interpreted in terms of the livre; that is 1·5 librae:

(a) Published mass in grams.   (b) Published rating.   (c) Hypothetical intended mass.

(a)	81·87	164·3	244·8	gm.
(b)	(1/6)	(1/3)	(1/2)	mina (=livre)
(c)	81·67	163·3	245	gm.

(The half-livre is inscribed 'Half a mina correct weight.')

Unmarked weights that can be rated in this scale include:

(a)	8·17	16·28; 16·42	40·85	81·9	163·9	gm.
(c)	8·16	16·33	40·83	81·6	163·3	gm.
=	(1/60)	(1/30)	(1/12)	(1/6)	(1/3)	livre

In terms of the livre, mina N is a double-mina; but its inscription reads:

> One true mina. Property of Marduk-Shar-ilani; copy of a weight which Nebuchadnezzar, king of Babylon, son of Nabopolassar, king of Babylon, made according to a weight of Shulgi, a former king.

Two weights that belonged to Shulgi are extant; the identity of their owner is established by these inscriptions:

To Nannar his lord has Shulgi the mighty man, king of Ur, king of the four lands, dedicated half a mina of correct weight.

Ten minas Shulgi.

The masses of these two weights (248 and 4986 gm.) are not quite in the scale of the livre: the lighter is nearly equal to 1 mark (= 8 oz.) troy (248·8 gm.), and the heavier is slightly more than 10 dimarks (4978 gm.) in the same scale. The mass of the troy dimark (symbol Tdm.) exceeds that of the livre by 1 part in 63. These Shulgi weights are the second and fifth in the following table:

(a)	41·5	248	496	1992	4986	14934	gm.
(b)		(1/2)		2	10		minas
(c)	41·5	248·8	497·7	1991	4978	14933	gm.
=	(1/12)	(1/2)	1	4	10	30	Tdm.

On the fourth weight in the above table, the inscribed rating is in terms of the double-mina (= 32 troy oz.); the inscription is:

(Cuneiform) Palace of Shalmaneser, king of the country.

Two minas of the country.

(Aramaic)   Two minas of the king.

The 30-mina weight is not marked; its mass is 40 lb. troy.

Shulgi's grandson, Gimil-Sin, owned the second weight in the following table; its mina $J = 504$ gm. is sometimes called the Babylonian light gold standard. It can be given a mass rating equivalent to 60 darics in terms of a daricweight of 8·4 gm., which is sometimes called the royal or light gold shekel. An important aspect of the significance of J is expressed by the equation $10J = 16$ Gm. = mass of 16 c. in. of gold; and this is the mass of the third weight in the table, but the inscription on it reads:

(Cuneiform) Five minas of the king.

(Aramaic)   Five minas of the country.

(a)	8·4	2511	5042	15061	60303	gm.
(b)		5	5	30		minas.
(c)	8·4	2520	5040	15120	60480	gm.
=	(1/60)	5	10	30	120	J.

The 30-mina weight is in the form of a swan,* and is inscribed:

30 minas true.   Palace of Eriba-Marduk, king of Babylon.

It belonged to Marduk-pal-iddina, a Chaldean prince of the house of Eriba-Marduk, tributary ruler in Bit Yakin. With military aid from Elam, he seized the throne of Babylon in 722 B.C. when Sargon II succeeded Shalmaneser V as king of Assyria. Driven out by Sargon in 710, he remained quiescent until Sennacherib came to the Assyrian throne in 705; then, sending envoys to Hezekiah (2 Kings xx. 12), he intrigued for a rising in Palestine to coincide with a second attack that gave him another short tenure of kingship. His biblical name is Merodach-Baladan (Isaiah xxxix. 1).

The heaviest weight in the above table is known as the 'Lion of Khorsabad'; it is in the Louvre.

Summarizing, these are the minas disclosed by this analysis:

1·5 Gm.	1·5 Gp.	1·5 Li.	16 Toz.	1·6 Gm.	1·5 lb.
7290	7350	7560	7680	7776	10500 gt.
472·5	476·28	490	497·78	504	680·4 gm.
	35				50 mass ratios.
		36			
		35		36	
		63	64		
			80	81	
15				16	

In the headings to the above table; it is necessary to recognize 1·5 Li. as 1·5 librae = livre, and 1·5 lb. as mina D. The mina rated 16 Toz. is 16 troy oz. = troy dimark, and the mina rated 1·6 Gm. is identified in the text by the symbol J. The minas rated 1·5 Gm. and 1·5 Gp. have equal significance in the sense that both can be regarded as equivalent to the mass of 1·5 c. in. of gold; but the ratios linking the former to 1·6 Gm. and the latter to 1·5 Li. justify the distinction that I have made between them.

This picture of ancient Babylonian weights is only so much of the whole as I have been able to assemble in an interlocking pattern with sufficient support for each of the standards to carry

---

* See page 8.

conviction: it derives importance from the names associated with it, and from the fact that it is in intelligible relationship to the standards that survived. Further research cannot fail to widen the field, but the assessment of intended values may become increasingly difficult; in any case it will require the greatest care. For example, there is in the Ashmolean Museum a limestone duck-weight * (from Erech) with a published mass of 2417 gm.:

Babylonian limestone duck weight from Erech. Unmarked. Hypothetical rating = 5 minas in terms of a mina = (16/15) lb. = k troy lb. Ashmolean Museum photo of exhibit 1912.1162.

although unmarked it is obviously in the 5-mina category; but what was its mina? Hitherto the nearest well-known standard was that called 'light gold' and here symbolized by J; but 5 of these minas would weigh 2520 gm. The livre offers a 5-mina standard of 2450 gm., but I suggest a 5-mina rating in terms of a mina of (16/15) lb. = k troy lb. for k = (35/27). This makes its intended mass 2419 gm., and shows the craftsmanship to be free from error. Alternative equivalent ratings are 5 × (80/81, livre, and 5 × (24/25) J. A talent of 60 such minas would weigh 64 lb., which is the mass of the Knossos octopus-weight.†

---

* No. 1912.1162.   † See IX(11).

V(7) ENTEMENA'S VASE

This fine example of the silversmith's art (*c.* 2400 B.C.) was found by de Sarzec during his excavation of ancient Lagash, at Tello: it is now in the Louvre. The silver vase, which is supported on a copper tripod base, is decorated with an engraved design in which the central device has been imaginatively described as the heraldic arms of the city; it shows a lion-headed eagle clutching the backs of two lions facing in opposite directions. An inscription records its dedication by Entemena to the god Ningirsu in his temple of Eninnu, during Dudu's high-priesthood: Entemena was the fifth governor at Lagash, during the Third Dynasty of Kish.

Thureau-Dangin * expressed the opinion that this vase probably was intended to be the niggin of 10 qa that is mentioned in tablets of the Sargon period, and he published his measurement of its volume as 4·71 litres. This I interpret as 1000 Sumerian c. shusi, which virtually identifies this qa with the troy pint. The arithmetic is:

$$\text{Volume of vase} = 4·71 \text{ litres} = 287·4 \text{ c. in. (Thureau-Dangin)}$$
$$= (6·6 \text{ in.})^3 = 1000 \text{ Sumerian c. shusi}$$
$$= (1/8) \quad ,, \quad \text{c. ft.}$$
$$\therefore \quad ,, \quad ,, \text{ qa} = 100 \text{ c. shusi} = (1/80) \quad ,, \quad ,,$$
$$28·8 \text{ c. in.} = \text{troy pint} = (1/60) \text{ English} \quad ,,$$

The Sumerian foot measures 1·1 ft., and its cube is 1·331 c. ft. = (4/3) c. ft. approx. ∴ (1/80) Sumerian c.ft. = (1/60) c.ft.

It is worth noting that the upper part of the body of this vase is hemispherical, and that the profile of the lower part corresponds to an arc having a radius equal to the diameter of the hemisphere; but struck from a point slightly below its base.

---

* *Zeitschrift für Assyriologie*, 1903, p. 24; *Journal Asiatique*, 1909, p. 912; *Revue d'Assyriologie*, 1921, p. 128.

ENTEMENA'S VASE

This silver vase (in the Louvre) is inscribed with the names of Entemena, governor of Lagash *c.* 2400 B.C., and of Dudu the high priest.   Hypothetical volume = (1/8) Sumerian c. ft. = 10 troy pints.

V(8) Babylonian Bricks

On one side of tablet YBC 7284 (Yale Babylonian collection) there are three lines of cuneiform that Neugebauer and Sachs translate to read:

One brick. What is its weight? Its weight is $8 + (1/3)$ minas.

On the other side is this cryptic assembly:

41 40   8 20   igi-gub-ba-bi 12   1 3 40

The editors believe that this brick was intended to measure $15 \times 10 \times 5 = 750$ c. shusi; which is equivalent to $(41/60^3) + (40/60^4)$ volume sar, and explains the meaning of the 41 40. Multiplying the volume by (12×60^2) gives 8; 20 * as its weight in sexagesimal notation. In this case, therefore, the meaning of 'igi-gub-ba-bi 12' is (as the editors explain) 'its density is 12×60^2 minas per sar.† The number 1 3 40 remains a mystery.

In terms of the Assyrian shusi of 0·648 in., the volume of this brick would be about 204 c. in.; and in terms of mina D, its mass would be exactly 12·5 lb. This represents a density of just under 1 oz./c. in.; which is a little less than that of the modern stock brick having a volume of 108 c. in., when rated 330 to the ton = 6·78 lb. each. At 1 oz./c. in., such a brick would weigh 6·75 lb. It appears, therefore, that the mina of this tablet could be mina D.

Other sizes of bricks identified from the texts, by Neugebauer and Sachs, include 18×12; 20×10; 20×20; 30×30; all in terms of the shusi.

---

* Following Neugebauer, I use here a semicolon to separate the whole number from the sexagesimal fraction. Thus, 8; 20 = $8 + (20/60)$; and 2, 3, 4; 20, 40 would mean $(2 \times 60^2) + (3 \times 60) + 4 + (20/60) + (40/60^2)$, and so on. Thureau-Dangin's numbers are accented in a manner similar to those used for angular measure. It should be understood that although it was the scribe's duty to separate the powers of 60, he had nothing equivalent to our cipher and decimal point with which to indicate place value: punctuation by the translator does not alter the sexagesimal figure sequence, but it does make it into a number.

† Neugebauer and Sachs use the word 'coefficient' as the general equivalent of 'igi-gub-ba,' because other tablets show it in association with long lists of numbers for use in practical mensuration.

V(9) MENSURATION OF LOGS

Among the cuneiform texts edited and translated by Neuge-bauer and Sachs, there is one dealing with the volumetric measure-ment of a log. Paraphrasing the published version, the problem and its solution are:

> If 1 kus is the circumference of a log, how big is it? You multiply (1/12) by (1/12) to give (1/144) and this by the factor 288 which gives 2 as the answer. Thus 2 sila is the cubic (ku-bu-ur) of the log.

This result is equivalent to writing:

> Let C = circumference, measured in kus.
> Then $2C^2$ sila = the cubic of the log.

Analysis of the calculation in the text shows that this sila was the cube of the linear gin ($= 216$ c. shusi); and that the cubic of the log was its volume in cubic gin per gin of length.   Thus:

$$\text{Sectional area} = (C^2/4\pi) \quad \text{sq. kus}$$
$$= (2C^2/25) \quad \text{,,} \quad \text{,,} \quad \text{for } \pi = (25/8)$$
$$= (1/144)\ (2C^2/25) \quad \text{,, gar}$$

$$\text{Sq. gar} = 60^2 = 144 \times 25 \text{ sq. gin}$$

$$\therefore \text{ Sectional area} = 2C^2 \text{ sq. gin for } \pi = (25/8)$$
$$\therefore \text{ Cubic of log} = 2C^2 \text{ c. gin per gin of length}$$
$$= 2C^2 \text{ sila}$$

Probably this sila was the trading unit on which the price was quoted: in terms of the Sumerian c. ft. of 8000 c. shusi rated (4/3) English c. ft., this sila = 0·036 c. ft. = 62·2 c. in.

This text is historically important as cuneiform evidence for the use of (25/8) for π.

V(10) METROLOGICAL RATIOS IN THE CUNEIFORM TEXTS

Isolated metrological ratios, in the cuneiform mathematical texts, collectively establish these principal units:

Gar (pole) = 12 kus (cubits); Cubit = 30 shusi.
Area sar = sq. gar = 144 sq. kus.
Vol. sar = gar².kus = 144 c. kus = 180000 vol. sila.
Vol. sila = linear gin³ = 216 c. shusi.
Sila of capacity; Mina of mass; Sar of bricks = 60 dozen.
Gin = (1/60)th of the gar, sar, sila, mina. Se = (1/180) gin.
Beru = 1800 gar; Bur = 1800 area sar; Gur = 300 capacity sila.

Problem texts, relating to the building of military ramps (for scaling city walls) and the digging of ditches, show lengths and widths in gar; heights and depths in kus; volumes in the mixed unit gar².kus, which is carried through the calculation. The unit of volume for logs and bricks, however, was the volume sila; representing the cube of the linear gin.

In terms of the Assyrian shusi of (8/9) digit = (k/2) in., the gar measures 16 remens = 20 Roman ft., and the bur of 1800 area sar is equal to 25 jugera; this means that the jugerum measures 72 sar = 72 × 144 Assyrian sq. cubits.

In terms of the Sumerian shusi of 0·66 in., the kus measures (1/10) English pole; and the jugerum measures 1 myriad Sumerian sq. cubits.

Problem texts support both interpretations. For example; two tablets (VAT 7848 and AO 6414),* published by Neugebauer and Sachs, contain texts relating to the sowing of seed barley at 36 sila per myriad sq. kus. If this sila is interpreted as a capacity for 2 mina D (= 3 lb.) of seed, and the kus as the Sumerian cubit of (1/10) English pole; then the rate of sowing is 108 lb. = 100 livres per jugerum. This is equivalent to 1 livre per sq. pole, or 3·2 bushels per acre with barley at 54 lb. (= 50 livres) per bushel; and this is within the range of modern farming practice. Moreover, the same rate also results from interpreting another text (VAT 8389) in terms of the Assyrian cubit. In this problem, a farmer sows 3 gur of seed per bur of area; which means 900 sila per 1800 sar, or mina D per sar (for sila = 2D as before); and so the seed sown on 25 jugera (= bur) is 1800D = 1800 × 25 unciae

---

* AO = Antiquités Orientales (Louvre).
VAT = Vorderasiatische Abteilung, Tontafeln. Staatliche Museen Berlin.

= 2500 × 18 unciae = 2500 livres.   Thus, the rate is 100 livres per jugerum, or 1 livre per sq. pole, as before.

In a tablet found at Nippur, the text gives 33 + (1/3) sila of seed for an area (called sq. gi) measuring 100 × 100 cubits: assuming this to be the jugerum, the rate of sowing is 1 lb. (instead of 1 livre) per sq. pole.   This particular text refers to 'the shusi of which 24 make the cubit for measuring the sq. gi'; on the above hypothesis, this shusi measured (1/24) Sumerian cubit.

Ancient Egyptian Measures and Weights

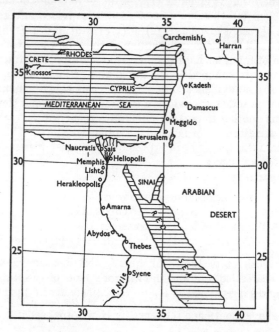

VI(1) EGYPTIAN LAND MEASUREMENT

ALTHOUGH the royal cubit was so called by Herodotus in his
description of Babylon, and although Oppert found evidence of its
use there; nevertheless, it has a predominantly Egyptian provenance through its association with the Great Pyramid and with the
Rhind Mathematical Papyrus.

Peet's translation of that document discloses a linear unit called khet, measuring 100 cubits, and a unit of area called setat that was 1 khet square. This, therefore, measured 1 myriad sq. cubits; but it is clear (from the context) that the setat should be visualized as 100 parallel strips each 1 cubit wide; and the name for such a unit was 'cubit of land.'

Thus, in problem 55, where an area of 3 setats is divided into 5 fields, the result is given as (1/2) setat + 10 cubits of land; and (in problem 54) where an area of 7 setats is divided into 10 fields, the size of each is given as (1/2) + (1/8) setat + 7 + (1/2) cubits of land.   All fractions in Egyptian arithmetic were expressed as the sum of fractions having unity as numerator; and fractions of the setat were confined to the dimidiated series (1/2), (1/4), etc.

There is no evidence that the setat was called a hundred of land; but 10 setats was called 'a thousand of land,' and Herodotus mentions another important unit in his reference (ii. 168) to the award of 12 arurae of land to Egyptian warriors.   There is no doubt that this should be interpreted as 12 setats; it is nearly equal to 13 Roman jugera.

Summarizing:

'A cubit of land' = 100 royal sq. cubits
Setat = 100 cubits of land
'A thousand of land' = 10 setats
'Warriors' land' = 12 setats = 13 jugera nearly

In English units, the royal square cubit measures 425 sq. in. to the nearest sq. in.; and the warrior's land of 12 setats would be 8·1 acres if reckoned as 10k = 12·96 (instead of 13) jugera.

The above-mentioned reference by Herodotus reads:

The warrior class in Egypt had certain special privileges . . . each man had 12 arurae of land assigned to him free of tax (the arura is a square of a hundred Egyptian cubits, the Egyptian cubit being the same length as the Samian).   All the warriors enjoyed this privilege together, but there were other advantages that came to each in rotation.   (Rawlinson's translation.)

According to the scribe Ahmose, the Rhind Mathematical Papyrus was copied by him during the thirty-third year of the

reign of the Sixteenth Dynasty Hyksos king Aauserre Apopi II.*
The original, he says, was written when Amenemhet III of the
Twelfth Dynasty was on the throne; and it was with this dynasty
(founded by the first Amenemhet *c.* 1991 B.C.) that Egypt entered
the Feudal Age, known as the Middle Kingdom, and came under
the rule of a succession of kings whose character and energy were
epitomized for the Greeks in the legendary Sesostris. It was, says
Herodotus, from the practice of this king in respect to the
measurement and assessment of land for taxation purposes, that
the Egyptians first learnt geometry; and it was from Egypt that
this knowledge ultimately passed into Greece. The name
Sesostris commemorates those of the three Senusrets, but the
tradition is that of the third Senusret and the third Amenemhet.
These were 'two of the greatest rulers that not only Egypt but even
the world has ever seen.' †

VI(2) THE ROYAL CUBIT

On documentary and other evidence, Griffith came to the con-
clusion that the square on the royal cubit was intended to be twice
the area of the square on the remen; and Petrie identified the
remen as a length of 20 digits. The royal cubit, therefore, can be
defined as $\sqrt{2}$ remen = $20\sqrt{2}$ digits = 20·62 in., for a digit of

The royal cubit as the semi-
diagonal in a square of 10
Roman ft. perimeter, and
as the radius in a circle of
10 Assyrian ft. circum-
ference.

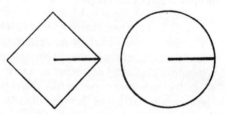

---

* See Bibliography (part 1: Peet). This torn papyrus was bought in
Luxor, by A. H. Rhind in 1858, and is now in the British Museum (No.
10057, 8): fragments, from the gap between the two parts, are in the possession of the Historical Society of New York. The original length of the
papyrus seems to have been about 216 in., and its width just under 13 in.
† H. R. Hall in the *Cambridge Ancient History*, vol. i, p. 302.

0·729 in. But, if 0·729 is expressed by $(9/16)k$; and $(35/27)$ is used for k, with $(99/70)$ for $\sqrt{2}$; the length of the royal cubit appears as $20 + (5/8) = 20·625$ in.; and 96 such cubits are equal to 10 English poles, or 100 Sumerian cubits.

Writing $20\sqrt{2} = 9\pi$ digits, gives the royal cubit a metric length of $(\pi/6)$ metre $= 523·6$ mm., for a digit of $(1/54)$ metre. Geometrically, this expresses the royal cubit as the sextant arc in a circle of 1 metre diameter.

Another geometric aspect is presented by the royal cubit regarded as radius in a circle of 10 Assyrian feet $= 100k$ in. circumference; its definition in this case is $(50k/\pi) = 20·6265$ in.

On the hypothesis that the digit is correctly defined (in accord with the evidence relating to the Greek and Roman feet) as $(1/54)$ metre, or 0·729 in., then the length of the royal cubit is defined by these formulae; and it can be used as a criterion of workmanship in monuments built to measurements in terms of this cubit. The most important of these is the Great Pyramid; in this, the side of the square base was intended to measure 440 such cubits. Surveys, made by Petrie in 1882, and by Cole (for the Egyptian Government) in 1925, gave lengths of 9068·8 and 9069·4 in. respectively; corresponding to 440 cubits of 20·612 in.

In the interior of this pyramid is the King's Chamber, measuring 20 by 10 cubits. Its measurement by Greaves (1637), Vyse (1837), Piazzi Smyth (1865), and Petrie (1882), gave values for the proportional cubit ranging from 20·5 to 20·66 in.*

Some extant royal cubit-rods are divided into 7 palms, one of which is longer than the rest: this extra length should be about a fifth of an inch. If the royal cubit is divided into $7 \times 4 = 28$ equal parts, then these are slightly longer than digits: conversely, a 7-palm cubit of 28 digits ($= 20·412$ in.) is shorter than the royal cubit by just over 1 per cent. Nevertheless, it seems probable that such a cubit was in existence, and that it was used by the architect of the great hall in the Temple of Karnak at Thebes. Such a cubit might seem to be more rational (in every sense of the term) than one of $20\sqrt{2}$ digits; but it would hardly be realistic on that account to doubt the reality of the royal cubit. Although

---

* This is an interesting example of variation in the recorded length of a constant object measured by different observers.

20√2 digits exceeds 28 digits by only 0·284 digit, the discrepancy in a length of 440 cubits is more than 7 ft. It may be said, therefore, that the evidence of the Great Pyramid alone is sufficient to establish beyond doubt * the existence of a royal cubit greater in length than 28 digits. Indeed, the 28-digit standard should, perhaps, be regarded as a debased form of 20√2 digits—corresponding to the use of (7/5) for √2—and not as one derived independently by the addition of a palm to the Roman cubit.

VI(3) HERODOTUS ON THE ROYAL CUBIT

In his description of the city of Babylon, Herodotus says, 'The royal cubit is longer by three finger-breadths than the common cubit.' † By common cubit, he means the Greek cubit rated 24 fingers: in effect, therefore, he defines the royal cubit as 27 Greek fingers in length. As the Greek finger measures (25/24) digits, Herodotus' definition of the royal cubit is equivalent to $27(25/24) = 9 \times (25/8)$ digits. Writing π for (25/8) gives this definition the significance of 9π digits = $(\pi/6)$ metre; which is my metric definition of the royal cubit.

An example of a royal cubit that can be interpreted as 27 Greek fingers is apparent on the Harmhab cubit-rod.

In Egypt, there was a version of the royal cubit 28 digits in length: the discrepancy between this length and 27 Greek fingers is the same as that between 28 and 27 × (25/24).

VI(4) PLINY ON THE ROYAL FOOT

In the sixth book of his *Natural History*, Pliny also gives measurements of the city of Babylon; but in terms of a foot that, he says, is 'three digits longer than ours.' ‡ As the Romans

---

* Note how the colossal size of the Great Pyramid facilitates these distinctions; and see further examples in VI(7).

† Herodotus, i. 178. 'Ο δὲ βασιλήιος πῆχυς τοῦ μετρίου ἐστὶ πήχεος μέζων τρισὶ δακτύλοισι (Oxford Classical Texts).

‡ Pliny, *Nat. Hist.*, vi. 26. 30. 'In singulos pedes ternis digitis mensura ampliore quam nostra.'

divided their foot into 16 digits, Pliny's definition of the royal foot makes it 19 digits in length; which is correct to the nearest digit.

Royal foot = (2/3) cubit = (2/3)20√2 digits = 18·8 digits.

VI(5) THE CUBIT OF THE NILOMETER

In Egypt there was a 6-palm cubit that slightly exceeded the length of the Roman cubit of 24 digits = 17·496 in.; and I think it originated as the side of a square that was half the area of the square on cubit A. On this hypothesis, its intended length was (1/2)(A√2) = 450 mm. = 17·71 in.

For convenience of reference, I call this standard the cubit of the Nilometer; because it appears to have been used as the unit of length in the scale on the ancient monument that indicated the water-level in the Nile. Hussey * refers to it thus:

> The cubit of the Nilometer was found by Le Père from the mean length of sixteen marked on that monument to be 239·7 lines in French measure or 17·742 inches.

In the Petrie collection there are two cubit-rods in this category; and one of them is inscribed with the names of Tutenkhamen and his wife Onkhesamen: it has a double scale. The published measurements of the palms and overall lengths are:

						Total
2·81	2·97	2·88	3·06	3·00	3·03	17·66 in.
2·85	2·96	2·86	3·05	3·01	3·06	17·79 in.
2·95	2·97	3·03	2·91	2·97	2·76	17·6 in.

There is good reason to suppose that a version of cubit A became the Palestinian cubit of √3 remen: by analogy, it is reasonable to assume that the semi-diagonal of its square acquired the significance represented by √(3/2) remen. This would have a square in (3/4) area ratio to the square on the royal cubit: moreover, as (3/4) = (36/48) and (36/49) = (6/7)², such a cubit would be roughly (6/7) royal cubit = 17·68 in. in length.

---

* Hussey (1836) gives his source as the *Mémoires sur l'Égypte pendant les Campagnes du Général Buonaparte*, vol. ii, pp. 32, 279.

VI(6) THE HARMHAB CUBIT-ROD

Among extant cubit-rods, one made in the reign of Harmhab is the best preserved and the most interesting: it was found at Memphis early in the nineteenth century, and is now in the Turin Museum; the measurements given here are based on those published by Jomard in 1822. On the diagram they are:

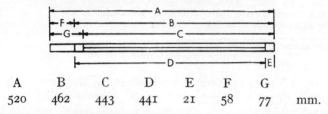

A	B	C	D	E	F	G	
520	462	443	441	21	58	77	mm.

It has a scale of 28 divisions that are not all equal: for example, part E is noticeably longer than the rest: thus, although there are 25 divisions in B its length is exactly 22E. Similarly, the three divisions of F are above the average, although shorter than E; and (G–F) stands on its own. The published measurements make B = 6G.

Lengths A, B, C, and D are in the following proportions:

A	:	B	:	C	:	D	
9	:	8					
		25	:	24			
		22			:	21	
520	:	462·2	:	443·5	:	441·2 mm.	(Proportionals)
520	:	462		443		441 mm.	(Published)

Hypothetically, the royal, Greek, and Roman cubits were represented by A, B, and C respectively; the lengths of these cubits are:

Royal = 523·6   Greek = 462·9   Roman = 444·4 mm.

If B is interpreted as the Greek cubit of 24 fingers = 25 digits = 462·9 mm., then A is a version of the royal cubit, represented by

THE HARMHAB CUBIT-ROD

Sketch (based on a drawing published by Lepsius in 1865) showing the
scale at each end. The lowest band is on a vertical edge, and the subdivisions continue through 5, 6, etc., to 16 parts. The middle pair of
bands are on a bevelled face. The inscriptions referring to Harmhab and
to Amen-em-opet are not shown.

27 Greek fingers (as defined by Herodotus), that was rated 28
digits in Egypt. Its published length of 520 mm. lies between
(28/54) metre and 27 × 0·76 in. Its (9/8) length ratio to the
Greek cubit implies that the square on the latter was considered
to be equal in area to that of a circle of royal cubit diameter. In
the Rhind Papyrus, a circle is squared by squaring eight-ninths of
its diameter; and by this convention the geometric significance of
$(1/2)\sqrt{\pi}$ was accorded to the rational ratio (8/9).

The length C must be regarded as 24 digits (= Roman cubit):
and if (B/D) = (22/21) really was intentional it may have been
$(\pi/3)$, for π = (22/7); but it will be noticed that the same significance attaches to (B/C) if (25/8) stands for π.

Harmhab is traditionally regarded as the founder of the Nineteenth Dynasty; but his relationship, if any, to his successor
Ramses I is unknown. On a statue in the temple of Horus at
Alabastronopolis (his home town) Harmhab records his rise from
official to pharaoh. In the reign of Ikhnaton (Amenhotep III)
he was an officer in the army: from the anarchy that followed the
death of Tutenkhamen he emerged, with the goodwill of the
priests of Amon, as the restorer of order; and his rule is said to
have been notable for the personal vigour with which he endeavoured to eradicate bribery and corruption.

In addition to the inscription that includes the cartouche of
Harmhab, there is another referring to a high official, possibly the
vizier, named Amen-em-opet.

E.N.A.

Two of the principal Pyramids at Gizeh.   On the right is the Great Pyramid built by Cheops; on the left is that built by Chephren.   Both pharaohs belonged to the Fourth Dynasty, *c.* 2650 B.C.   All the Pyramids have been despoiled of their casing stones that gave them smooth exteriors.   The head of the Sphinx, supposed to have been carved in the likeness of Chephren, is just visible above the sand in the direction of his Pyramid.

VI(7) THE GEOMETRY OF THE PYRAMIDS

When Vyse, excavating at the base of the Great Pyramid in 1837, uncovered some of the casing stones, the angle of slope θ became known to the modern world for the first time.   Omitting seconds, the first published value was $\theta = 51° 51'$; in 1883, Petrie (in a report printed with the aid of a grant from the Royal Society) published $\theta = 51° 52'$.   The cotangents of these angles are:

$$\text{Cot } \theta = \cot 51° 51' = 0.78551$$
$$0.785398 \ldots = (\pi/4)$$
$$\text{,,} \quad = \cot 51° 52' = 0.78504$$

The Rhind Mathematical Papyrus shows that the Egyptians called the vertical height pr-m-ws (from which the Greeks may have derived their word for pyramid) and used skd, in the sense that we use cot θ, to define the nature of the design: the question, therefore, is whether the Great Pyramid was intended to represent cot $\theta = (\pi/4)$; and if so, what was this intended to imply? *

Writing S = base side and H = height; then cot θ = (S/2H). And if cot $\theta = (\pi/4)$, then (S/2H) = (22/28); or (S/H) = (440/280), for $\pi = (22/7)$. In the light of Petrie's measurements, the appropriate ratings for the base side and height are 440 and 280 royal cubits. Conversely, if these measurements are accepted, then it must be admitted that they disclose in *clear numbers* the ratio (22/7) as a value for π. Moreover, if (in any pyramid) cot $\theta = (\pi/4)$, then it is a geometric fact that the perimeter of the square base is equal in length to the circumference of a circle described with the height as radius; this could be regarded as a geometric theme underlying the design of this pyramid. The relationship is expressed by writing:

$$(S/2H) = (\pi/4) \qquad \therefore \ 2\pi H = 4S \ \dagger$$

If the above interpretation of the geometry is correct, its execution in clear numbers predetermined the great size of this pyramid; for the only alternative to $4 \times 7 = 28$ cubits was 280 cubits for the height.

As, by tradition, the Greeks believed that they had acquired their first interest in geometry from the Egyptians, it is not only permissible but necessary to look for geometrical evidence in Egypt and especially to examine the pyramids in this light because they are the monuments of antiquity most likely to exhibit it. In the light of the measurements of the Great Pyramid it seems reasonable to suppose that the Egyptian geometers had investigated the properties of the circle and believed π to be $3 + (1/7)$. If this is true it is a matter of historical importance because it means that other values in common use, such as 3 and (25/8), were not the limit of current knowledge but conventions of greater convenience. Also it becomes legitimate to accept evidence implying that useful functions of π were interpreted

---

* This question was raised by Taylor, in 1859.
† Petrie noted this relationship.

numerically by the most suitable approximations without regard to constancy in the deduced value for π itself.   For example, the formula for calculating the area of a circle (as given in the Rhind Papyrus) is to square eight-ninths of the diameter, which corresponds to $\pi = (256/81)$.

The Great Pyramid was built for Khufu—known to the Greeks as Cheops—but he was not the first to indulge in that remarkable

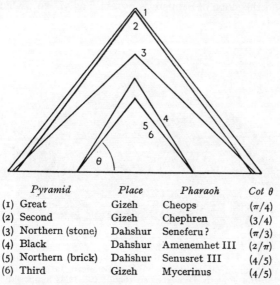

Pyramid	Place	Pharaoh	Cot θ
(1) Great	Gizeh	Cheops	$(\pi/4)$
(2) Second	Gizeh	Chephren	$(3/4)$
(3) Northern (stone)	Dahshur	Seneferu ?	$(\pi/3)$
(4) Black	Dahshur	Amenemhet III	$(2/\pi)$
(5) Northern (brick)	Dahshur	Senusret III	$(4/5)$
(6) Third	Gizeh	Mycerinus	$(4/5)$

form of mortuary architecture.   His father Seneferu, founder of the Fourth Dynasty, may have been responsible for three.   One of these (at Dahshur) was measured by Perring, who published its slope as $43°$ $36'$ which suggests cot $43°$ $31' = (\pi/3)$.   The published length of its base side is 719.5 ft.; if this is read as 720 ft. $= 400$ UU, the intended height appears as 200 royal cubits and the value of cot θ is given by the ratio of the units (UU/Rc) $= (\pi/3)$ when $(100k/6) = 21.6$ in. is written for UU and $(50k/\pi)$ for Rc=royal cubit.

Also at Dahshur is the Black Pyramid built for Amenemhet III of the Twelfth Dynasty: published by Perring as $57°$ $20'$, its slope might have been intended for 57° 31′, giving $(\pi/4)$ for (S/H) and $(2/\pi)$ for cot θ.   The base side is published as 342·5 ft., which implies an intended length of 200 royal cubits = 343·7 ft.   This measurement (200 Rc.) probably was the intended height of another pyramid at Dahshur, that built for the equally famous Twelfth Dynasty king Senusret III: it was the same size as that built at Gizeh for Mycerinus of the Fourth Dynasty.   Perring published the slope of Senusret's pyramid as 51° 20′; which gives cot $\theta = (4/5)$.

E.N.A.

The Bent Pyramid (at Dahshur) built by Seneferu, father of Cheops and founder of the Fourth Dynasty, c. 2690 B.C.   It is supposed that the reduced slope of the upper part was due to some urgent need to complete the construction.   The discovery (1947) of Seneferu's name in this Pyramid makes it unlikely that he built (as formerly supposed) the northern stone Pyramid at Dahshur.

The Mycerinus Pyramid at Gizeh is the smallest of the group of three that is dominated by the Great Pyramid: the second of these was built for Chephren, and its measurements indicate $(3/4)$ as the probable intended value of cot θ.   In the Rhind Papyrus there is a problem in which cot $\theta = (3/4)$ is expressed as 5 palms + 1 finger per cubit: in terms of a cubit rated 7 palms (= 28 fingers), the ratio is $(21/28) = (3/4)$.   It will be observed that 5 palms + 2 fingers per cubit would give cot $\theta = (22/28) =$

($\pi/4$), for $\pi = (22/7)$.   Another problem in the papyrus requires the pupil to find the 'batter' of a hypothetical pyramid having a base side of 360, and a height of 250, cubits.   Peet's translation of the solution is:

> You are to take a half of 360; it becomes 180.   You are to reckon with 250 to find 180; result $(1/2) + (1/5) + (1/50)$ of a cubit.   The cubit being 7 palms, you are to multiply by 7; its batter is $5 + (1/25)$ palms.

The meaning of this is:

$$\text{Cot } \theta = (1/2)(S/H) = (180/250) = (1/2) + (1/5) + (1/50)$$
$$= 5 + (1/25) \text{ palms per cubit}$$
$$= \cot 54° \ 15'$$

This problem derives particular interest from the fact that this angle is the same as that published by Perring for the lower slope of the Bent Pyramid at Dahshur: it was built for Seneferu, and is so called because the upper part has a much reduced angle.

The Step Pyramid

This pyramid, designed by Imhotep for Zoser (probable founder of the Third Dynasty *c.* 2816 B.C.), was built at Saccara, the necropolis of Memphis.   Its flat summit forms the sixth step above the rectangular base; but archaeological evidence shows this final form to have been an enlargement of a design having a summit four steps above a square base.

Perring's measurements of the height, step by step from bottom to top, show that successive steps diminish by either 21 or 20 in.; which suggests an intended constant decrease of a royal cubit of 20·625 in.   Also, it implies that the royal cubit was the unit of measurement; and it will be noticed that the total recorded height of the first five steps is 2055 in., corresponding to 100 cubits of 20·55 in.   The published measurements in inches, and my interpretation of them in terms of the royal cubit, are:

$$452 + 431 + 411 + 391 + 370 + 350 = 2405 \text{ in.}$$
$$22 + 21 + 20 + 19 + 18 + 17 = 117 \text{ cubits}$$

If this height has any particular significance, I do not recognize it; but the number 117 itself is intriguing, for it appears again (2000 years later) in the number of ingots that Croesus sent as a present to the Oracle at Delphi.*

The length and breadth of the rectangular base of this pyramid are published as 4727 and 4214 in.; which implies 230 and 204 royal cubits respectively: they also suggest an intended area of 47000 royal sq. cubits, or 20 million sq. in.   The product of this area by the height is virtually 550 myriad c. cubits: and the sum of the three linear measurements is 551 cubits.

A wall enclosed the site on which this Pyramid was built; and its boundary stones, inscribed with the names of the king and two of his daughters, were found by Lauer.   His measurements show the length and breadth as 544 and 277 metres; and if each is increased by a metre, the area would be very close to 60×96^2 royal sq. cubits = 60 jugera: probably this was its intended significance.   The corresponding length × breadth in terms of the royal cubit would be $1040 \times 530 = 551200$ royal sq. cubits; but the true area of 60 jugera is 552960 in terms of this unit.   The design seems to have incorporated the figure sequence 55 as a theme.

The entrance passage of the Great Pyramid

A downwardly sloping entrance passage facing north is a feature of several of the pyramids: that of the Great Pyramid has an angle of 26° 41′ to the horizontal, and (as the latitude of Gizeh is 30°) points 3° 19′ below the pole of the heavens.   An observer standing about 63 ft. inside the passage would see 7° of the celestial polar region framed in the opening, and would be able to note the culmination of any circumpolar star whose orbit crossed that area.

Sir John Herschel reported to Vyse that:

The passage may be said to have pointed directly at α Draconis at its inferior culmination, at which moment its altitude above the horizon of Gizeh would have been 27° 9′.   Four thousand years ago, the present pole star α Ursae Minoris could by no

---

* VIII(6)

possibility have been seen . . . at its lower culmination it was only 7° above the horizon of Gizeh.

The entrance passages in the Chephren and Mycerinus pyramids (as published by Vyse) are 25° 55′ and 26° 2′ respectively.

VI(8) THE GREAT HALL IN THE TEMPLE OF KARNAK AT THEBES

Egyptian power and magnificence in the time of the Empire were nowhere better exemplified than in the architecture of the Temple of Karnak at Thebes.   Here was the 'greatest colonnaded hall ever erected by man'—338 ft. wide by 170 ft. deep, with a floor area about equal to that of the cathedral of Notre-Dame in Paris, and this was only a single room in the temple.*   The nave (79 ft. high) was divided into three aisles by two rows of colossal columns that are still standing.

Evidently the width of this hall was intended to be twice its depth, and this latter measurement probably was intended to be 100 cubits of 28 digits = 170·1 ft.; the floor area, therefore, was 2 myriad sq. cubits = 2 setats.   This cubit of 28 digits was a version of the royal cubit, and can be said to represent its rating of $20\sqrt{2}$ digits, if 1·4 is used for $\sqrt{2}$.   The height of the nave, published as 79 ft. = 948 in., is only a fraction of an inch in excess of 1300 digits = 947·7 in.

It was during the period of foreign occupation by the Hyksos (who provided the kings of the Fifteenth and Sixteenth Dynasties) that Thebes became a focal point of intermittent revolt by tributary Egyptian princes of the south.   In particular, there was Sekenenre III, Taa; he married Princess Aahotep and was succeeded by their three sons: first by Kamose, then by Senekhtenre, and finally by Ahmose, who liberated his country from the foreign invaders.

---

* *Ancient Times,* by J. H. Breasted (1935), p. 108.

VI(9) QUEEN HATSHEPSUT'S OBELISK

Immediately behind the Great Hall—and, therefore, in the midst of the buildings forming the Temple of Karnak—stood two obelisks erected by Queen Hatshepsut in celebration of her royal jubilee. 'These obelisks were the tallest shafts ever erected in Egypt up to that time, being ninety-seven and a half feet high and weighing three hundred and fifty tons each. One of them still stands and it is possible that the queen set up two more pairs, making six in all.' *

This published height of 97·5 ft. suggests 97·2 ft. = 1600 digits = 80 Egyptian remens = 50 Assyrian cubits = 100 Roman ft.

Makere-Hatshepsut, the first Great Queen in world history, was the great-grand-daughter of Ahmose the Liberator, who founded the Eighteenth Dynasty after driving the Hyksos out of the land about 1574 B.C. Her father Thutmose I succeeded Amenhotep I in 1534, and married three times; in 1505 she succeeded her elder half-brother Thutmose II, but there is some doubt whether his relatively insignificant reign was really independent of hers. In 1495 she married her younger half-brother, but he seems to have derived little satisfaction from his position as prince consort to such a dominant personality; for when he came to rule alone as Thutmose III (1485–1441) he very thoroughly obliterated all the monumental inscriptions of her reign: time, however, revealed her obelisks when his enclosing brickwork collapsed.

The obelisks, like the Fourth Dynasty Pyramids, show the pharaohs' characteristic disregard of engineering difficulties: consider the time required to quarry these monoliths by using a stone in the hand to pound the surrounding rock to dust. Would they have succeeded in erecting the 1100-ton monster that lies unfinished † in the granite quarry at the First Cataract of the Nile? Its published length is 137 ft., which suggests an intended length of 80 royal cubits; the proportional cubit in this case being 20·58 in.

---

* J. H. Breasted in the *Cambridge Ancient History*, vol. ii, p. 65 (1924).
† Abandoned when flaws revealed themselves in the shaft.

VI(10) EGYPTIAN WEIGHTS

Documentary references and the evidence of extant weights suggest that the basic Egyptian system of mass was: 1 sep = 10 debens = 100 kites. Relevant inscriptions on weights are: 'Five kites. Treasury of On'; 'Uahabra. 1 deben. 5 khnp kites'; 'The sep of Uahabramernet.'

Egyptian half-deben. Inscribed '5 kites. Treasury of On.' Mass =(1/10) lb. British Museum photograph of exhibit 33871.

Relatively few of the many extant ancient Egyptian stone weights have rating marks; but this evidence is sufficient to indicate the importance of the kite as a rating unit, and to show that there were many different kites: this diversity probably reflects the pre-eminence of Egypt as a centre to which traders came from all parts of the known world. These are some of the kites that I have derived from published masses and ratings:

gt.		gt.	
69·4	= (1/70) Gm.	230·4	= (1/25) troy lb.
140	= (1/50) lb.	240	= (1/24) ,, ,,
189	= (1/40) livre	243	= (1/20) Gm.
210	= (1/24) libra	245	= (1/20) Gp.
218·75	= (1/32) lb.	291·6	= (1/24) lb.

Examples of ancient Egyptian stone weights in which these kites are the rating units include the following selection from about eighty published weights that I have analysed in this way.

SOME EGYPTIAN MARKED WEIGHTS

Published: (*a*) Mass; (*b*) Rating mark.
Hypothetical: (*c*) Intended mass; (*d*) kite.

	(1)	(2)	(3)	(4)	(5)	(6)	(7)	(8)
(*a*)	139	209	436	481	569	698	1380	1382 gt.
(*b*)	(1/2)	1	1	2	3	5	6	3
(*c*)	140	210	437·5	480	569	700	1382	1382
(*d*)	280	210	437·5	240	189	140	230·4	460·8

	(9)	(10)	(11)	(12)	(13)	(14)	(15)	(16)
(*a*)	1398	1404	1460	1471	2080	2188	14571	14700 gt.
(*b*)	10	5; (½)	5; 1	6	30	10	60	60
(*c*)	1400	1400	1458	1470	2083	2187	14580	14700
(*d*)	140	280	291·6	245	69·4	218·75	243	245

Three of these kites are more readily recognizable as:

210 gt. = half-uncia; 218·75 = half-oz.; 240 = troy half-oz.

Weight (2), for example, is half an uncia in mass but its rating
mark is 1; similarly, weight (4) has a mass of 1 oz. troy and its
rating mark is 2; but weight (3) is 1 oz. averdepois in mass and its
rating mark is 1.   Petrie mentions another 1-oz. weight (published
mass = 439 gt.), found in the tomb of a First Dynasty queen
named Sma-nebui, but it has no rating mark.   Weight (14) has a
mass of 5 oz., but is rated 10: it belonged to a Tenth Dynasty
priest named Ampi, and is now in the Berlin Museum.

Weights 1, 6, 9, and 10 are rated in terms of a kite of 140 gt. =
(1/50) lb. = (1/36) libra, or its double.   There are two extant
examples of (6); one in the British Museum, and one in a private
collection; both are inscribed '5 kites. Treasury of On.' and can
be regarded as half-debens.   The proportional 10-kite deben
would have a mass of (1/5) lb.; and the proportional sep of 10 such
debens would have a mass of 2 lb.   In the Petrie collection there
is an unmarked weight (4714) with a published mass of 11215 gt.
that can be rated 8 debens (= 11200 gt.) in this scale; its owner
inscribed himself:

Hereditary prince, purifier in the temple of Ptah, sam priest,
high priest of Memphis.   Hora.

On weight (9), which is a 10-kite deben, the rating mark shows the Egyptian numeral ∩ = 10 in the middle of ten strokes, thus: ΙΙΙΙΙ ∩ ΙΙΙΙΙ. On weight (10), the same mass is associated with the ratings 5 and (1/2); which imply the use of a double-kite, and the existence of a double-deben. This variation in rating is characteristic of ancient metrology: it appears again in weights (7) and (8); the former being rated in terms of (1/25) troy lb. as a kite, and the latter in terms of its double. It should be noted, however, that the ratings on these weights are 6 and 3; the proportional 10-kite deben would be (2/5) troy lb. In weight (4), as already mentioned, the kite is (1/24) troy lb. = (1/2) troy oz.

Egyptian sep (No. 15 in the table) mentioned on page 88. Inscribed 60. Mass = 3 Gm. British Museum photograph of exhibit 23067.

Weight (11), now in the Vienna Museum, is of some historical interest: it belonged to Psammeticus II Apries (588–566 B.C.) who succeeded Necho II in the Saite line forming the Twenty-sixth Dynasty in Egypt. He was the pharaoh who persuaded Zedekiah of Judah to ignore the warning of Jeremiah and to revolt against Nebuchadnezzar II, who had appointed him to the governorship

(2 Kings xxiv. 17) after deposing Jehoiachin. The weight is inscribed 'Uahabra. 1 deben. 5 khnp kites'; a ratio that suggests the use of (1/24) lb. as a double-kite. In weight (10), the rating unit is (1/25) lb., and is also a double-kite.

Metrologically, the most interesting weights in the above table are the two seps (15) and (16); their masses are 3 Gm. and 3 Gp. respectively, and both weights are marked 60 in terms of a kite corresponding to (1/20) c. in. of gold in mass: the same kite (245 gt. in this case) forms the basis of the rating mark 6 on weight (12). Another weight in this scale is (4355) in the Petrie catalogue; it is a rectangular block of green-veined marble, with a cambered upper surface, on which a column of hieroglyphs describes its owner as:

> Hereditary prince, royal seal bearer, sole companion, keeper of the seal, Herfu living again.

The published mass is 19621 gt.; which can be called 4 Gp. = 19600 gt.

VI(11) THE GOLD-SIGN WEIGHTS

In the dictionary used by students of Egyptology there is the word nub, meaning gold; its hieroglyph is: ⌒⌒⌒⌒

This sign is engraved on several Egyptian stone weights, including one inscribed 'The Goldsmith Homeri': possibly, therefore, it may have been used as a certificate of accuracy by bankers and others who dealt in this metal. As the following table shows, it signified neither a constant mass nor a constant unit.

Published: (a) mass, (b) rating. Hypothetical: (c) mass, (d) kite.

	(1)	(2)	(3)	(4)	(5)	(6)	(7)	(8)	(9)
(a)	73	97	194·7	323	784	1021	1183	1277	2012
(b)	(1/4)	(1/2)	1	3	4	5	6	6	8, 10
(c)	72·9	97·2	194·4	324	784	1021	1181	1280	2016
(d)	291·6	194·4	194·4	108	196	201·4	196·8	213·3	201·6
									252

The hypothetical kites are:

gt.		gt.	
108	= (1/45) Gm.	196·8 = (1/32) funt	
194·4	= (1/25) ,,	201·6 = (1/25) libra	
196	= (1/25) Gp.	213·3 = (1/36) troy dimark	
204·1	= (1/24) ,,	291·6 = (1/24) lb.	
252	= (1/20) libra = (1/30) livre = (1/60) mina N		

One of the above weights (5) is engraved with the cartouche of Amenemhet III: he succeeded Senusret III in the Twelfth Dynasty (*c.* 1991–1778 B.C.). Amenhotep I, son of Ahmose the Liberator, was the owner of weight (6): he was the second king in the Eighteenth Dynasty, which ruled Egypt (*c.* 1574–1315 B.C.) after the expulsion of the Hyksos. On weight (7) the inscription is 'Thotmes. Beloved of Ptah': it is in the Louvre, and so is weight (5).

Amenhotep's gold-sign weight. British Museum photograph of 38546.

The masses of the two royal weights are (4/25) and (5/24) Gp. respectively, and that of weight (4) is (1/15) Gm. A small gold bar, found in the tomb of the First Dynasty king Aha, is mentioned by Petrie; its published mass (216 gt.) is exactly two-thirds of the mass of (4), but is more readily recognized as (1/25) tower lb.

Goldsmith Homeri's weight has no rating mark, but its published mass (853 gt.) is two-thirds that of (8); and an appropriate rating would be 4 kites, in terms of a kite rated (1/36) troy dimark.*

The lightest of the gold-sign weights (1) has the heaviest kite = (1/24) lb. = (2/3) oz.; the mass of the weight itself is only (1/6) oz.   The heaviest weight in the list (9) has two ratings, 10 and 8;

* Dimark = 2 marks = 16 oz.; a name coined for use when pound means 12 oz.

the proportional kites are (1/25) libra, and (1/20) libra = (1/30) livre = (1/60) mina N.

It is relevant to remark that the kite rated (1/24) lb. is equivalent to (1/25)(3/2) Gm., when the gold-mina is rated (25/36) lb.: the analysis of Babylonian marked weights discloses 1·5 Gm. as a rating unit.

VI(12) CAPACITY RATIOS IN THE RHIND PAPYRUS

In the Rhind Mathematical Papyrus there are six problems dealing with large corn bins, and they teach the pupil to multiply the volume (measured in cubic cubits) by (3/2) in order to express the capacity in khar of 20 hekat rating. This hekat contained 10 hins, and this hin was rated 32 ro. Thus:

Ro		Hin		Hekat		Khar		C. cubit
9600	=	300	=	30	=	(3/2)	=	1

In terms of a royal cubit of just over 20·6 in., the khar appears as the cube of 18 in.; that is to say, its volume is (1/8) c. yard. In terms of a cubit of nearly 20·63 in., the khar is equal to 96 litres: its true value is between these narrow limits. This interpretation is supported by published measurements of an Egyptian bronze bowl (No. 27) and a bronze vase (No. 8) in the Petrie collection at University College, London:

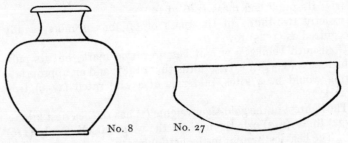

No. 8   No. 27

No. 27. 546·5 c. in. = (1/16) c. cubit; for khar = (1/8) c. yd.
  ,,   8. 366·2   ,,   = (1/16) khar   ,,   ,,   = 96 litres

The hin proportional to a khar rated (1/8) c. yd. has a volume equal to (81/80) troy pint, in terms of a troy pint defined as (1/60) c. ft., but the only marked vases in the Petrie collection show the troy pint itself as the hin. One of these has three calibration rings, and the published value at the level of the middle ring is 28·9 c. in.: the other vase is marked (1/8), and the published volume is 3·61 c. in. One-eighth of a troy pint, rated 28·8 c. in. = (1/60) c. ft., would be 3·6 c. in.

That there were other hins is apparent from the volume and rating of an alabaster vase in the British Museum. It is marked 8 + (1/6) hins, and its volume has been published * as 269·57

Inscription, 8 + (1/6) hins, on the alabaster jar illustrated in the frontispiece. The fraction (1/6) is on the left; the word h-n on the right.

c. in., which implies a hin close to 33 c. in. Also in this collection is an unmarked alabaster vase of 271·19 c. in.; and if the mean of these similar volumes is taken to be a hekat of 10 hins, the proportional hin would be 27 c. in. And 216 of these would be equal to a khar rated (1/8) c. yd.: this, being a sexagesimal ratio, may have been the intended relationship.

The hekat was an official corn measure in Egypt, and its sub-multiples in the dimidiated series (1/2) to (1/64) were always written in symbols now known as Horus-eye notation.

This descriptive name is derived from the fact (first perceived in modern times by Möller) that these symbols can be arranged to form a conventionalized outline of a human eye. It is their separate use as fragments of this assembly, however, that causes the reference to Horus: it is an allusion to the story that he had an eye torn to pieces in his fight with Seth. In one of the spells (Ch. 17) of the *Book of the Dead* the deceased declares: 'I have

---

* Measured by W. Airy and published in the *Proceedings of the Institution of Civil Engineers* (1908–9), vol. clxxvii, part 3, p. 164. See frontispiece.

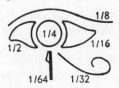

completed the eye after it had been injured on the day of combat between the two rivals.' It has been suggested that the reference to completion may be intended to have its arithmetical parallel in the fact that the dimidiated series $(1/2) + (1/4)$, etc., needs the repetition of the last fraction to make unity.

Hekat	$(1/2)$	$(1/4)$	$(1/8)$	$(1/16)$	$(1/32)$	$(1/64)$,
Ro	160	80	40	20	10	5
Symbol	◁	○	⌒	▷	◡	╵

VI(13) TWO GRADUATED CONICAL MEDICINE CUPS

In the Museum at Cairo there are two Eighteenth Dynasty graduated conical cups;[*] one is made of silver, the other of bronze. On the silver cup there are 10 calibration rings: the highest, which is well below the brim, is marked hin, and the next is marked dja; then come the $(1/2)$ hin and the $(1/2)$ dja. The remaining fractions are all in terms of the dja, and the process of dimidiation is carried down to $(1/128)$. The bronze cup is broken just below the $(1/2)$ dja, and the silver cup is damaged near the bottom; but linear measurements of both have been published [†] and from calculations based on these I deduce that both cups were designed with a common cone angle $\theta = 27°$. Further calculations show the (bronze/silver) dja volume ratio to be $(3/2)$ and suggest that their intended water-weights were 18 unciae = 1 livre and 12 unciae = 1 libra respectively. Similarly, the silver (hin/dja) volume ratio is seen to be $(25/24)(3/2)$, which gives 18 oz. as the water-weight of the silver hin.

Independently measured quantities of water may have been used to establish the calibration levels, but there are features of the geometry that suggest calculation. For example, the volume of the bronze dja is exactly 50 times the volume

[*] These cups are catalogued (Nos. 5513, 5514) as Eighteenth Dynasty; but some authorities assign them to the Graeco-Roman period.
[†] By A. Lucas and Alan Rowe; in *Annales du Service Antiquités de l'Égypte*, vol. xl (1939). See XIV(3).

q of the invisible (but, in calculations, crucially important) cone below the base, and this bronze q is two-thirds the volume of the silver q: this is the inverse of the dja volume ratio. Moreover, if the silver q is regarded as a ro of (9/10) c. in. in the Egyptian scale of capacities revealed by the Rhind Papyrus, then the proportional hin of 32 such ro has a volume of 28·8 c. in. = (1/60) c. ft., which is my volumetric rating for the troy pint. Proportionally, the silver dja is a volume of 20 c. in. and its water-weight at a density of

The silver cup.

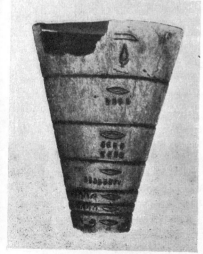

The bronze cup.

0·036 lb./c. in. is 0·72 lb. = 1 libra.

If, as suggested by Lucas and Rowe, these are medicine cups, then the fluid dose represented by (1/128) dja was: silver = 2 scruples; bronze = 1 drachm, both in the French pre - metric apothecaries' scale. This was in (63/64) mass ratio to the units of corresponding name in the English troy scale. Diluted to the dja level, the medicinal draught was about half a pint.

VII

Palestinian and Biblical Units of Measurement

VII(1) THE PALESTINIAN CUBIT

EPIPHANIUS was Bishop of Constantia in Cyprus and already seventy-seven when, in A.D. 392, he wrote a treatise on weights and measures from material compiled some years previously at the instigation of the Emperor Valentinian II. The oldest surviving fragments are written in Greek, but the only complete texts are two Syriac versions in the British Museum; and translations of these, by J. Elmer Dean, are the basis of the following interpretation of the metrology.

The crucial equation, derived from a somewhat involved statement, is that the Palestinian jugon of first-class land (also called a koraean) contained 30 sataeans; and was equivalent to 13 Roman jugera. There is a reference to Egyptian practice in land measurement that justifies the interpretation of 13 jugera as 12 setats, for the setat was equal to the square on 100, and the jugerum to the square on 96, royal cubits:

$$\therefore \text{(Setat/Jugerum) area ratio} = (25/24)^2 = (13/12) \text{ approx.}$$

Two measuring poles were in use: the rod of 5 cubits and the akaina of (4/3) rod. There were also two land-units. The smaller measured 20 rods by 20 cubits, making an area of 2000 Palestinian sq. cubits: the larger (called sataean) measured 20 akaina by 20 cubits, so that its area was (4/3) × 2000 sq. cubits. Thus, the jugon of 30 sataeans contained 8 myriad Palestinian sq. cubits = 12 myriad Egyptian sq. cubits.

$$\therefore \text{(Palestinian/Egyptian) sq. cubit area ratio} = (3/2)$$
$$\therefore \quad \text{,,} \qquad \text{,,} \quad \text{cubit length ratio} \quad = \sqrt{(3/2)}$$
$$\text{Royal cubit} = \sqrt{2} \text{ remen} \therefore \text{ Palestinian cubit} = \sqrt{3} \text{ remen}$$
$$= 25 \cdot 25 \text{ in.}$$

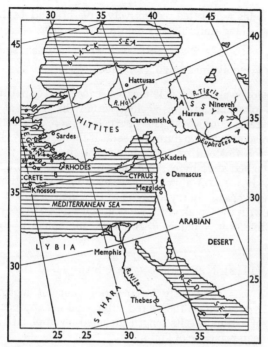

The terms in which Epiphanius describes the Palestinian cubit
are the same as those in which Ezekiel (xliii. 13) refers to the cubit
of the altar; that is to say it is the cubit and handbreadth. The
above analysis shows the cubit of reference to be the royal cubit;
but it is possible that the Palestinian cubit's relationship to it was
forced in order to give the convenient (3/2) sq. cubit area ratio. If
so, then the cubit of the altar (commonly called the sacred cubit)
may have had an independent origin; and if this was cubit A,
the Palestinian field of 1 myriad such sq. cubits would have been
the geodetic acre.

There were two fields of first-class land:

Palestinian = (1/8) jugon = 1 myriad Palestinian sq. cubits
Egyptian = (1/6) „ = 2 „ royal „ „
= 2 setats

These fields were rated respectively 5 plethra and 5 sataeans in terms of the land-units already mentioned, namely:

Plethron = 20 rods × 20 cubits = 2000 Palestinian sq. cubits
Sataean = 20 akainas × 20 ,, = 4000 royal   ,,   ,,

The jugon was called a koraean because it required a kor of seed.   A plausible rate of sowing, within the range of modern farming practice for wheat, is 2·5 bushels per acre at 64 lb./bushel. This corresponds to 100 livres per setat if the jugon is reckoned as 10k = 12·96 (instead of 13) jugera.   On this basis, the kor of wheat weighed 1200 livres; and the proportional ephah, or bath (Ezekiel xlv. 11, 14) appears as a capacity for 120 livres = 60 mina N of wheat.   In my analysis of the 'molten sea,' I identify the bath with the Greek c. ft., as a capacity for 60 livres of water. In the Bible: kor = homer = 10 ephas, and ephah = bath.

VII(2) Newton on the Sacred Cubit

Newton was sufficiently interested in ancient metrology to write a dissertation on the sacred cubit; and his approach to the problem of estimating its length included an application of the Archimedean principle of limits:

> To the description of the Temple belongs the knowledge of the Sacred Cubit; to the understanding of which, the knowledge of the cubits of the different nations will be conducive.   That the Sacred Cubit was very large, appears from the Jewish Calamus or Reed, which contained but six of these cubits; and from the antiquity of this cubit since Noah measured the Ark with it. However, it is not to be magnified in such a manner that the vulgar cubit (which in the time of Moses was called 'the cubit of a man': Deuteronomy iii. 11) should much exceed the cubit of a tall man.   But we shall circumscribe these cubits in narrower limits in the following manner.
>
> The stature of the human body, according to the Talmudists, contains about 3 cubits from the feet to the head.   Now the ordinary stature of men, when they are barefoot, is greater than 5 Roman feet and less than 6 Roman feet.   Take a third part of this and the vulgar cubit will be more than 20 unciae and less

than 24 unciae of the Roman foot; and consequently the Sacred Cubit will be more than 24 unciae and less than 28 + (4/5) unciae of the same foot.*

This upper limit is in the same ratio to 24, as 24 is to 20; and by continuing to apply this principle of limits to other evidence that he considered to be relevant, Newton reduced the range of choice to 25·27 to 25·79 Roman in. From this, he infers 'that the Sacred Cubit of Moses was equal to 25 unciae of the Roman foot and (6/10) uncia.' This is equivalent to 24·88 in.; but it is interesting to note that his upper limit was 25·067 in., or virtually cubit A.

VII(3) The 'Molten Sea'

When Solomon had been king over all Israel for three years, he began to build the house of the Lord; and, in the south-east corner of the court of the priests, he set the great brass water-tank called the 'molten sea.' The description in 1 Kings vii includes the following measurements:

> Ten cubits from the one brim to the other: it was round all about, and his height was five cubits: and a line of thirty cubits did compass it round about. And it was an handbreadth thick; it contained two thousand baths.

In 2 Chronicles iv, the capacity is given as 3000 baths. These measurements can apply to a hemisphere and a cylinder having a depth equal to the radius, as stated in both descriptions.

Hiram, the brassfounder, cast the bowl 'in the plain of Jordan; in the clay ground between Succoth and Zarthan.' Probably it was cast bottom up, with the core of the mould forming a hemispherical mound in the centre of a pit; and Hiram encircled the base of this core with a measuring cord when checking its size. I interpret the text to mean that this measurement was 30 cubits and, therefore, that the *internal* radius of the bowl was $(30/2\pi)$ cubits; and its volume as a hemisphere $2(\pi/3)r^3 = 2000$ baths.

---

* Published in 1736 by Thomas Birch; who said that it was 'translated from the Latin of Sir Isaac Newton, not yet published.' It is included in the book containing Greaves's miscellaneous works.

This shows the bath to be a capacity for 60 livres of water when the cubit is rated $(2/\pi)$ metre. The arithmetic is:

$$\text{Volume} = 2(\pi/3)(30/2\pi)^3 = 2000(\pi/3)(3/2\pi)^3 \text{ cubic cubits}$$
$$= 2000 \text{ baths} \qquad (\text{I Kings vii})$$
$$\therefore \text{Bath} = (\pi/3)(3/2\pi)^3 = (9/8\pi^2) \qquad \text{cubic cubits}$$
$$= (9/\pi^5) \text{ c. metre, for cubit} = (2/\pi) \text{ metre} = \text{cubit A}$$
$$= (1/3\cdot24)^3 \text{ c. metre, for } \pi^5 = 31 \times (800/81)$$
$$= \text{Greek c. ft.} = \text{capacity for 60 livres of water.}$$

This simultaneous identification of the cubit with cubit A, and the bath with the cube of the Greek foot, is of particular metrological significance; because both linear units are defined geodetically, and the association confers reality on cubit A as the original length of the sacred cubit mentioned by Ezekiel (xliii. 13). Also it supports the hypothesis that a talent of 60 livres, derived from the water-weight of the Greek cubic foot, was the origin of the livre itself and of the ancient metrology of mass.

As the external diameter was 10 cubits; the ratio of the external to the internal radius was $5/(30/2\pi) = (\pi/3)$, and the thickness was $(\pi - 3)$ cubit = handbreadth = 3·58 in.

VII(4) BIBLICAL MEASURES

Ezekiel xliii. 13. 'These are the measures of the altar after the cubits. The cubit is a cubit and an handbreadth.' xl. 5. 'A measuring reed of six cubits long by the cubit and an handbreadth.'

Ezekiel xlv. 11. 'The ephah and the bath shall be one measure, that the bath may contain the tenth part of an homer, and the ephah the tenth part of an homer.' xlv. 14. 'The cor, which is an homer of ten baths.'

Exodus xvi. 36. 'An omer is the tenth part of an ephah.'

Ezekiel xlv. 12. 'And the shekel shall be twenty gerahs: twenty shekels, five and twenty shekels, fifteen shekels shall be your maneh.'

$$\therefore \text{Mina} = 20 + 25 + 15 = 60 \text{ shekels}$$

Exodus xxxviii. 25, 26. 'And the silver of them that were numbered of the congregation was 100 talents and 1775 shekels, after the shekel of the sanctuary. A bekah for every man; that is half a shekel, after the shekel of the sanctuary, for every one that went to be numbered, from twenty years old and upward, for 603550 men.'

$$603550 \text{ half-shekels} = 300000 + 1775 \text{ shekels}$$
$$\therefore \quad 300000 \text{ shekels} = 100 \text{ talents}$$
$$\therefore \quad 3000 \quad ,, \quad = 1 \text{ talent}$$

Biblical references to monetary values:

Genesis xxiii. 15, 16. 'My lord, hearken unto me: the land is worth 400 shekels in silver. . . . And Abraham weighed to Ephron the silver, which he had named in the audience of the sons of Heth, 400 shekels of silver, current money with the merchant.'

Exodus xxi. 32. 'If the ox shall push a manservant or a maidservant; he shall give unto their master 30 shekels of silver, and the ox shall be stoned.'

1 *Chronicles* xxi. 25. 'So David gave to Ornan for the place 600 shekels of gold by weight.'

1 *Kings* x. 29. 'And a chariot came up and went out of Egypt for 600 shekels of silver, and a horse for 150.'

Hypothetically, the mass of the Hebrew shekel was (1/2) oz. averdepois, and its value in silver might be rated 10 pence by analogy with the 20-pennyweight rating of the ounce troy: or it might be rated half a crown, in terms of the English silver penny when its weight had fallen to one-third of a pennyweight; but neither takes account of relative purchasing power. The horse at 150 shekels offers some comparison of value.

Biblical references to integrity in measurement:

Proverbs xi. 1. 'A false balance is abomination to the Lord: but a just weight is his delight.' xvi. 11. 'A just weight and balance are the Lord's: all the weights of the bag are his work.' xx. 10. 'Divers weights and divers measures, both of them are alike abomination to the Lord.'

Deuteronomy xxv. 13–15. 'Thou shalt not have in thy bag

divers weights, a great and a small. Thou shalt not have in thine house divers measures, a great and a small. But thou shalt have a perfect and just weight, a perfect and just measure shalt thou have.'

Leviticus xix. 35, 36. 'Ye shall do no unrighteousness in judgment, in meteyard, in weight, or in measure. Just balances, just weights, a just ephah, and a just hin, shall ye have.'

Ezekiel xlv. 10. 'Ye shall have just balances, and a just ephah, and a just bath.'

Micah vi. 11. 'Shall I count them pure with the wicked balances, and with the bag of deceitful weights?'

Hosea xii. 7. 'He is a merchant, the balances of deceit are in his hand.'

Note the custom of carrying private weights in a bag when travelling on business: this would be necessary on account of the different standards in different places, and as a precaution against fraud.

These condemnations of false weighing and appeals for just measurement were reiterated by the rulers of all nations, and echo even in the preamble to the Act that embodies our present standards: and nowhere is this moral philosophy aspect of metrology more forcibly expressed than in the Anglo-Saxon laws:

Ethelred. In nomine domini, anno dominicae incarnationis M.VIII. This is the ordinance that the King of the English and both the ecclesiastical and lay witan have chosen and advised. . . . (24) And let fraudulent deeds and hateful illegalities be earnestly shunned; that is false weights and wrongful measures, and lying witnesses and shameful fightings.

It will be realized that the special significance of these quotations, in this context, lies in the fact that they certify the existence of true standards.

VII(5) HEBREW METROLOGY
(According to the International Critical Tables)

Sacred mina	=	60 shekels	Talmudist mina	=	25 shekels
,, talent	=	3000 ,,	,, talent	=	1500 ,,
		= 50 S. minas			= 60 T. minas

I.C.T.		*Interpretation*
Sacred mina =	850 gm.	30 oz. ∴ Shekel = (1/2) oz.
,, bath =	29·376 litres	64·8 lb. = 60 livres of water
Talmudist ,, =	21·42 ,,	Talmudist talent of water
Sacred cubit =	0·64 metre	25·2 in. = 2·16 Roman ft.
Talmudist ,, =	0·555 ,,	21·87 ,, = (3/2) remen

The Hebrew sacred bath and the Greek cubic foot are given as equal capacities in these tables of ancient measures, but the sources are not stated.

VII(6) EPIPHANIUS ON THE MEASUREMENT OF A LOG

Palestinian timber merchants used π^{-1} c. cubit as a trading unit, if their system of mensuration is correctly reflected in a formula for the volume of a cylindrical log given by Epiphanius. The true volume of a cylinder of length L and circumference C, both in cubits, is $(LC^2/4\pi)$ c. cubits; or $(LF^2/4\pi) \times 24^2$ c. cubits, if the circumference measures F fingers at 24 to the cubit. In the formula given by Epiphanius, π is omitted from the denominator and so $(1/\pi)$ becomes a measure of the unit in terms of which the measurement of volume is calculated. The form in which the denominator appears is 12×192, which is equivalent to 4×24^2. The relevant passage in Dean's translation of the text is:

This cubit has 24 fingers in the measure, if the cubit is a linear measure. . . . The measure of a piece of timber is taken from the circumference of the timber. For example, if you wind a cord about a piece of timber and it is found that there are in it 72 fingers, then you multiply the 72 fingers by 72 again, which makes 5184 fingers. You divide these again by 12 and there are 432 fingers. . . . You take the length of such a piece of timber. If it be 10 cubits, you multiply the 432 lepta by these 10 and there are 4320 lepta. Then you divide these by 192. . . . So there are, in a piece of timber that is 72 fingers in circumference and 10 cubits long, 22 and a half cubits.

In terms of a cubit of 25·25 in., the volume of π^{-1} c. cubit is nearly 3 c. ft.

VIII

Metrological Aspects of Money

VIII(1) MONEY BY WEIGHT AND TALE

GOLD and silver were used as bullion in remote antiquity, but merchant bankers took the first step towards its conversion into specie when they put private marks on pieces of which they knew the weight and had confidence in the purity.

> In the early documents from Cappadocia, dating from 2250 B.C., it is frequently stated of sums of money that they have to be paid in money 'of my seal' or 'of your seal.' In those of the Sargonid period 720–620 B.C. it is frequently agreed that so many minas are to be paid in minas of Carchemish; less often in minas of the king: and there is strong ground for believing that all these specifications are of the same type, namely, that the money was marked in a certain way.*

Perhaps the potential convenience of small pieces of uniform money, that could be accepted individually at sight without weighing, had been apparent long before the earliest documentary reference to their manufacture; this is an inscription in which Sennacherib (whose reign began in 705 B.C.), referring to the casting of some very large effigies, says:

> According to the command of the god I fashioned moulds of clay and poured bronze therein, as in casting half-shekel pieces, and I completed their construction.†

Whether these little moulds had any device that was reproduced on the metal is unknown; but the castings must have been

---

* 'A Pre-Greek Coinage in the Near East,' by Sidney Smith. *Numismatic Chronicle* (1922), p. 176.
† An inscription published (*Cuneiform Texts from Babylonian Tablets*) by L. W. King and quoted by Sidney Smith, ibid., p. 177.

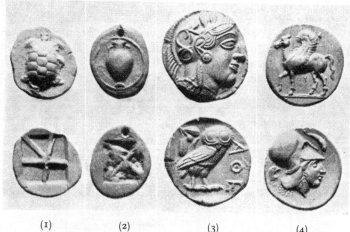

(1)   (2)   (3)   (4)

Some early Greek coins in the Ashmolean Museum (Heberden Coin Room)

(1) Aegina: didrachm.	Obv. Turtle.	Rev. Incuse square.		
(2) Athens: ,,	,, Amphora.	,, ,, ,,		
(3) Athens: tetradrachm.	,, Athena.	,, Owl.		
(4) Corinth: stater.	,, Pegasus.	,, Pallas.		

recognizable as half-shekel pieces, and presumably they were accepted as such without question.

The practice of making coins by stamping small pieces of metal, is supposed to have originated in one or other of the cities founded on the Asiatic coast by emigrants who fled from Greece when the Dorian invasion terminated the Mycenaean period: the earliest specimens are made of electrum (a natural alloy of gold and silver that was plentiful in Asia Minor), but they show only the mark of the punch used to force the face of the blank into the smooth recess of the anvil die. Coinage in the modern sense of the term began when the anvil die was engraved to reproduce on the face of the piece some device that identified its place of origin. A turtle was the badge that identified the coinage of Aegina; the winged horse Pegasus distinguished that of Corinth.

The Attic coinage introduced by Solon displayed an amphora;

but later Athenian types reflect the recrudescence of political strife between the merchants of the shore, then led by Megacles, and the farmers of the plain under Lycurgus. When Peisistratus founded a third party of the hills, he also started a new numismatic fashion by impressing a device on both sides of his coins; Athena's head and an owl decorated obverse and reverse respectively.

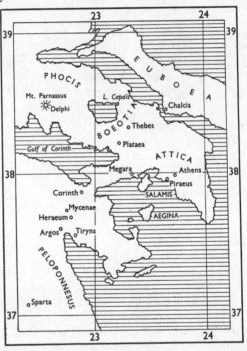

Although the didrachm and the tetradrachm were the principal denominations in general use; the drachma and its fractions down to its forty-eighth part were also coined, and these latter were so small that they were carried in the mouth by folk who had much need of them as cash. The sixth part of the drachma was called obol, and both names perpetuated the terminology of an earlier currency in which six iron spits (obeloi) made a real handful (drachma). In general, the metrological aspect of numismatics is outside the scope of this report, but some reference to the physical evidence relating to early Greek coinage is essential.

VIII(2) PHEIDON'S CURRENCY REFORM

Greek tradition attributed the introduction of the Aeginetan silver coinage to Pheidon,* King of Argos. Hitherto, thin iron rods had been used as currency; and a bundle containing 180 such spits was excavated on the site of the Argive Heraeum, by the American School at Athens. It must have been the actual dedicatory offering in commemoration of the reform: the tradition of it (recorded by Heracleides of Pontus in the fourth century B.C., as preserved in Orion's *Etymologicum*) is quoted by Seltman † thus:

> Pheidon the Argive first of all men struck coins in Aegina, and having issued coins, he removed the spits and dedicated them to Argive Hera.

The bundle.

A solid iron bar was found with the bundle: its published mass is 73000 gm. (bundle = 72540 gm.); and if this is assumed to have been a multiple of some established standard, the interpretation of intended mass that seems most plausible is:

The bar.

---

* In numismatic circles, doubt has been expressed about Pheidon's association with this reform; the issue is chronological.
† *Greek Coins*, by Charles Seltman (1933), p. 33.

Bar = 150 livres = 73500 gm.

∴ Bundle + bar = 300 livres
　　　　　　　 = 6 Babylonian talents of 50 livres *
　　　　　　　 = 5 talents of 60 livres, corresponding to the
　　　　　　　　　 mass of 5 Greek c. ft. of water †

∴ 　　　　Spit = (300/360) livre = 15 unciae = (9/10) lb.

∴ 　　Handful = 6 × 15 unciae = 5 livres

Extant evidence relating to early Aeginetan silver coinage includes 58 didrachms in the British Museum: the frequency table of these specimens is:

185	186	187	188	189	190	191	192	193	gt.
3	5	4	14	5	16	3	6	2	

If the proportional drachm is compared with the handful of iron spits on the hypothesis that they were of equal value as currency, then (on the basis of a 5-livre rating for the iron) the most rational interpretation of the silver evidence suggests a didrachm of (1/40) livre = 189 gt.; giving a (silver/iron) value ratio of 5 × 80 = 400, as deduced by Seltman.

Nothing in this evidence points clearly to any particular silver mina: if 60 drachms counted as such, the hypothetical mass was (3/4) livre; if 50, then (5/8) livre. It should not be overlooked, however, that the iron spit was already in the mina category of mass; and this standard may have remained in Aeginetan metrology as a mina of 15 unciae = (5/4) libra. The presence of a weight of this mass in Russian metrology suggests that it was an ancient standard.

VIII(3) PHEIDONIAN MEASURES

Tradition, mentioned by Herodotus (vi. 127) and by others, also associated Pheidon's name with measures and, therefore, it seems reasonable to look for some evidence of his standard of length in that of the spits found at the Argive Heraeum: this length (of the bar) has been published as 119 cm., which suggests:

---

* See VIII(7). 　　　　　　　　　† See I(10).

118·518 cm. = (64/54) metre = 64 digits
= 4 Roman ft. = 48 Roman in.
= French aune (ell)

Hypothetically, therefore, the spit of 15 unciae mass was intended to weigh (5/16) uncia per Roman inch of length; and in the analysis of Solon's currency reform, I suggest that (5/16) uncia = (3/10) oz. was the mass of the new Attic didrachm. This association also suggests that the Roman inch, of (4/3) digit = 0·972 in., was the Pheidonian standard (rather than the digit of 0·729 in.); and that the length of his spit may have been rated by the number 4, in terms of the unit subsequently known to history as the Roman foot. This hypothetical Greek usage of the Roman foot finds support in the presence of the sculptured impress of a foot of this length on a marble monument (of metrological import) now in the Ashmolean Museum; * for the marble itself is Greek (c. 450 B.C.), and is believed to have come from the west coast of Asia Minor.

VIII(4) SOLON'S CURRENCY REFORM

In 594 B.C., social distress among the Athenians reached the climax described by Aristotle: †

The many were in slavery to the few, the people arose against the upper classes. The strife was bitter . . . until at last, by common agreement, they chose Solon to be mediator and Archon, and committed the whole constitution into his hands. . . . Under him the mina, which previously contained 70, was filled out, so as now to contain 100 drachms; whereas the ancient standard coin was a didrachm.‡

If, hypothetically, this 'ancient standard coin' was a didrachm

---

* See IX(7).

† *Constitution of the Athenians.* Passages from chapters 5 and 10 are here placed in juxtaposition: the much-debated reference to the currency reform reads:

ἡ μνᾶ πρότερον ἔχουσα σταθμὸν ἐβδομήκοντα δραχμὰς ἀνεπληρώθη ταῖς ἑκατόν· ἦν δ' ὁ ἀρχαῖος χαρακτὴρ δίδραχμον.

‡ Seltman's translation in *Greek Coins*, p. 44.

equivalent in mass to the Aeginetan didrachm, then the mina of 70 proportional drachms weighed (7/8) livre. And if, hypothetically, Solon re-rated this mina at 100 drachms; then the intended mass of his new drachm was (7/800) livre = 66·15 gt.: and the proportional didrachm weighed 132 gt., to the nearest grain.

But, if the mina was an Attic metrological standard having an earlier origin than the currency; and if that currency was Aeginetan in mass; then there might have been a fractional discrepancy in the rating 70. And it seems relevant to remark that a mina of 15 oz. could be regarded as having the same metrological significance as one of (7/8) livre; for their mass ratio of (125/126) is the same as that of the gold-mina to the gold-pound. In terms of this hypothetical mina of 15 oz., Solon's new drachm weighed (3/20) oz. = (5/32) uncia = 65·625 gt.; and the proportional didrachm had an intended mass of 131 gt., to the nearest grain.

In these frequency tables, summarizing the weights of individual coins published by Seltman, the peak of the didrachm favours the lighter, while that of the tetradrachm supports the heavier, interpretation.

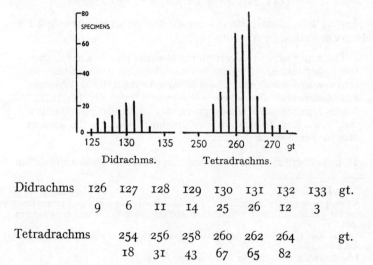

Didrachms.   Tetradrachms.

Didrachms	126	127	128	129	130	131	132	133	gt.
	9	6	11	14	25	26	12	3	

Tetradrachms		254	256	258	260	262	264		gt.
		18	31	43	67	65	82		

In combination, these tables imply that the above interpretation of Aristotle's comment is plausible; and they justify recourse to later evidence for an indication of present probability. In this wider field, the mass of the Attic drachms may be compared with that of the zolotnik in Russian metrology: originally this was exactly $(3/20)$ oz. $= (5/32)$ uncia if, as is highly probable, the pood originated as 50 librae $= 36$ lb. On this hypothesis there were 100 new Attic drachms in a 15-oz. Attic mina, and 96 in a hypothetical 15-uncia Aeginetan mina corresponding to the mass of the iron spit: in Russian metrology, the funt is rated 96 zolotniks; and the name of this small weight has a derivation meaning gold.

Another direction in which the mass of Solon's drachm survived, or reappeared, was in the Arabic weight (called mithkal) used as the standard of reference for the mass of the gold dinar in the Mohammadan monetary system. Weights made of glass were used to check the masses of the coins; and the peak frequency of those in the British Museum indicates a proportional mithkal of 66 gt., to the nearest grain: but it should be noted that $(3/20)$ oz. $= 65 \cdot 625$ gt. would also be recorded as 66 gt., to the nearest grain.

Presumably, some convenient rate of exchange was established when the new coinage was introduced, and Aristotle's numbers 70 and 100 suggest the ratio $(3 \times 70)/(2 \times 100) = 1 \cdot 05$: this could imply that Solon popularized the change among the rich by offering 3 new didrachms for 2 of the old, at a cost to the state of 5 per cent in bullion. For the poor perhaps he placed the new didrachm on a par with the old, for the discharge of debt.

VIII(5) THE CORINTHIAN 'COLTS'

At Corinth, the silver staters were nicknamed 'colts,' in reference to the winged horse Pegasus embossed on them. The following table (based on those assigned by Ravel * to the period

---

* *Les Poulains de Corinthe*, by O. E. Ravel (1936): 'A cause de ce petit cheval ailé ces monnaies étaient dans le peuple connues sous le nom de πῶλοι ou 'poulains,' p. 8. The weights are published in grams.

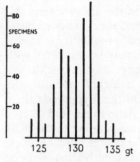

549–440 B.C.) shows a peak frequency corresponding to that of the Athenian reformed currency.

126	127	128	129	gt.
1	17	31	25	

130	131	132	133	134
31	61	74	34	10

In the Corinthian system, the stater was divided into 3 drachms; and the mass of this drachm was (1/10) oz., if the mina is interpreted as 15 oz.

VIII(6) CROESUS'S ELECTRUM INGOTS

Having, for two years, mourned the death of Atys—his favourite son, accidentally slain at the boar hunt by Adrastus—Croesus awoke to the Persian menace and sent a mission to consult the oracle at Delphi. Included among the magnificent gifts that his ambassadors delivered was the present of bullion that Herodotus describes thus:

> When the sacrifice was ended, the king melted down a vast quantity of gold and ran it into ingots, making them six palms long, three palms broad, and one palm in thickness. The number of the ingots was one hundred and seventeen, four being of refined gold in weight two talents and a half, the others of pale gold and in weight two talents. (Rawlinson's translation, i. 50.)

Croesus reigned as the last king of Lydia from 560 B.C. until he was captured by Cyrus of Persia in 546 or 545 B.C.; and Herodotus (born in 484 B.C.) probably was on his travels between 460 and 450 B.C. When he heard the story of these ingots, therefore, it is likely that less than a century had elapsed since they were cast.

Undoubtedly, the 'pale gold' was electrum; that is to say, 'native argentiferous gold containing from 20 to 50 per cent of silver.'* As Herodotus does not differentiate in size, his

---

* O.E.D.

(electrum/gold) ingot weight ratio implies that the density ratio also was (4/5), and this·would require a silver content of 30 per cent by weight.   It appears, therefore, that the gold in the 113 electrum ingots weighed 113 × 2 × (7/10) = 158 talents; and this with the four refined ingots made 168 talents of gold in all.

In the light of the evidence relating to the gold-mina, it seems reasonable to interpret the size of the ingot in terms of a palm of 3 in. = (1/4) ft., and the density of gold as 1200 lb./c. ft.   Thus, the four gold ingots had a total volume of roughly (9/8) c. ft.; and their combined weight of 10 talents can be interpreted as equivalent to 1350 lb. = 1250 livres, which implies a talent of 125 livres. The total gold represented by 168 talents, therefore, was about 21000 livres, or just over 10 tons.

If this talent of 125 livres is rated 60 minas, the proportional mina weighs 36 oz.: for comparison, mina N weighs 36 unciae. In terms of the livre, mina N is a double-mina: similarly, a mina of 36 oz. would be a double-mina in terms of the obsolete Russian bezmen; its mass is (3/2) mina D.

The number of ingots was 117, the same as the number of royal cubits in the height of the Step Pyramid.

VIII(7) THE PERSIAN TRIBUTE

In the year 521 B.C., a most co-operative neighing on the part of his horse, backed by the sporting assent of his fellow conspirators, enabled Darius son of Hystaspes peacefully to occupy the vacant throne of Persia; and there, by the aid of adequate finance and management, he maintained his authority for thirty-six years.

His revenue came in the form of tribute paid by twenty satraps appointed to govern the provinces into which he divided the empire.   They were assessed, of course; Herodotus gives a list of the amounts, and also explains (iii. 89) that

> such as brought their tribute in silver were ordered to pay according to the Babylonian talent; while the Euboic was the standard measure for such as brought gold.   Now the Babylonian talent contains 70 Euboic minae. (Rawlinson's translation.)

Only the tribute from India was paid in gold,* and Herodotus gives this as 360 Euboic talents. The total silver assessment of the other 19 satrapies was 7740 Babylonian talents; but as 140 talents were earmarked for the maintenance of certain cavalry, the total revenue actually remitted to Darius was:

360 Euboic talents of gold +

(7740 — 140) = 7600 Babylonian talents of silver.

Herodotus then says:

If the Babylonian money here spoken of be reduced to the Euboic scale, it will make nine thousand five hundred and forty such talents; and if the gold be reckoned at thirteen times the worth of silver, the Indian gold-dust will come to four thousand six hundred and eighty talents. Add these two amounts together, and the whole revenue which came into Darius year by year will be found to be in Euboic money fourteen thousand five hundred and sixty talents; not to mention parts of a talent. (Rawlinson's translation, iii. 95.)

It will be noticed that the sum of the two numbers, as given by Herodotus, is not equal to his stated total. Thus:

Herodotus	*Herodotus*
9540 + 4680 = 14220	14560 Euboic talents of silver

In a footnote, Rawlinson speaks of a double error quite impossible to rectify; but it is obvious that the 4680 Euboic talents of silver is a correct valuation of the 360 Euboic talents of gold, for a (gold/silver) value ratio of 13, and it is reasonable to suppose that its subtraction from the total of 14560 talents might give the correct equivalent, in the Euboic scale, of the 7600 Babylonian talents of silver. The result is:

$$(14560 - 4680) \ = \ 9880 \ = \ 1 \cdot 3 \times (7600)$$
$$\therefore \ 14560 \ = \ 13 \times (360) \ + \ 1 \cdot 3 \times (7600)$$

---

* Enlivening history with anecdote, Herodotus tells how this gold-dust was collected from remote desert sand-heaps thrown up by 'great ants, in size somewhat less than dogs but bigger than foxes.' ·They chased the miners!

This result, if assumed to be correct, gives 9880 as the appropriate correction for 9540; and shows at a glance that the (Babylonian/Euboic) talent value ratio (= mass ratio) was 1·3.

As this is one-tenth of the (gold/silver) value ratio, it is evident that the rate of exchange was 10 Babylonian talents of silver for 1 Euboic talent of gold.

The simplest way of summarizing the Persian tribute, therefore, is:

7600 Babylonian talents of silver are worth

760 Euboic talents of gold

Add the Indian tribute of   360   ,,   ,,   ,,   ,,

Total   1120   ,,   ,,   ,,   ,,

which are worth (13 × 1120) = 14560   ,,   ,,   ,, silver

This is the number given by Herodotus, but he adds the remark not to mention parts of a talent,' which implies that no whole number could be correct.

He may mean that the (gold/silver) value ratio was not exactly 13, or that the (Babylonian/Euboic) talent mass ratio was not exactly 1·3: this says the same thing in another way, but permits the discrepancy to be illustrated more readily by reference to the fact that he also says 'the Babylonian talent contains 70 Euboic minae.' This implies that the mass rating of the Euboic talent, in its own scale, was:

Euboic talent = (70/1·3) = 53·86 Euboic minae

Certainly this is wrong; but if the (gold/silver) value ratio (given by Herodotus as 13) is changed to the sexagesimal number $10k = 12·96$, in the form 10 × (35/27); and the (Babylonian/Euboic) talent mass ratio is changed correspondingly from 1·3 to $k = 1·296$, in the form (35/27); then the 70 Euboic minae in the Babylonian talent are seen to be proportional to 54 such minae in the Euboic talent. Thus:

Euboic talent = (70/k) = 54 Euboic minae, for k = (35/27)

In modern editions of the Greek text, the 9540 is emended to read 9880; and it is evident that this is recognized as equivalent to 1·3 × 7600, because there is a further emendation whereby

Herodotus is made to say that the Babylonian talent contained 78 instead of 70 Euboic minae.   This number, 78, is the product of 60 × 1·3, and expresses the editor's assumption that the numerical rating of the Euboic talent must have been 60 Euboic minae.

How and Wells in their commentary on these modern texts, which read (ὀκτὼ καὶ) ἑβδομήκοντα μνέας, say: 'Reizke's (Mommsen, *Röm. Münz.*) conjecture to add ὀκτὼ καὶ is usually accepted. The MSS. here, however, have only ἑβδομήκοντα and this reading is as old as Pollux.'

There can be little doubt, therefore, that Herodotus wrote and meant 70:* emendation is unnecessary if my interpretation is accepted; but I agree that the Euboic talent probably had a sexagesimal rating of some kind, and I suggest that this may have been 60 Gm. (= 54 tower lb.) representing the mass of 60 c. in. of gold.

This hypothesis implies that the Euboic mina was a prototype tower pound, and that the Babylonian talent weighed 70 tower lb. = 54 lb. averdepois = 50 livres.

A sexagesimal rating for the Babylonian talent is 60 Q, where Q = (9/10) lb. = 15 unciae; this Q, I think, was the prototype Russian funt.

VIII(8) The Persian Daric

Darius obtained 32 per cent of his revenue as gold dust from his satrap in India: the rest was paid in silver, by the governors of the nineteen other provinces.   'The Great King,' says Herodotus,

stores away the tribute which he receives after this fashion; he melts it down and, while it is in a fluid state, runs it into earthen vessels, which are afterwards removed, leaving the

---

* Rawlinson's translation conforms to the manuscripts in retaining 70; but the calculation that he made in a footnote was based on the total assessment, and not on the net revenue.   He did not deduct the 140 talents earmarked for the maintenance of certain cavalry and, therefore, did not suggest 9880 as a necessary correction of 9540; he concluded with the remark that 'it is impossible to reconcile Herodotus's numbers and equally impossible to say where the mistake lies. . . .

metal in a solid mass. When the money is wanted, he coins as much of this bullion as the occasion requires. (Rawlinson's translation, iii. 96.)

In the Persian coinage, the gold daric was worth 20 silver sigloi. Estimates of the mass of the daric have been published by the following authorities:

	gm.
Weisbach	8·34
Viedebanth	8·35
Petrie	8·37
Hultsch	8·40
Thureau-Dangin	8·40
Brandis	8·50

The Daric.

The Siglos.

Evidence of extant coins suggests that the intended mass of the siglos was 84 gt. (= half the mass of the Croesus stater), which makes the intended mass of the daric 8·4 gm. = 129·6 gt. if the (gold/silver) value ratio was $10k = 12·96$, as deduced from my analysis of the Persian tribute.

A mina of 60 such darics would weigh 504 gm. (for which I use the symbol J); and 10 such minas are equal to 16 Gm. Works of reference, and labels on ancient weights in museums, often give 8·4 gm. as the Babylonian 'light or royal gold standard shekel'; but it is only one of many ancient standards of mass. The associated 'light silver standard' is commonly published as 11·2 gm.: this is $(4/3) \times 8·4$ gm.; and implies a (gold/silver) value ratio of $(40/3)$ if the daric of 8·4 gm. was worth 10 silver staters of 11·2 gm., but this mass does not seem to be supported by the evidence of either weights or coinage. Twice the mass of the siglos, that is to say 2×84 gt. $= (1/30)$ libra $= 10·87$ gm., would be more appropriate.

It is important to note that 8·4 gm. $= (1/54)$ lb., for lb. $= 453·6$ gm.

IX

The Metrologies of Greece and Rome

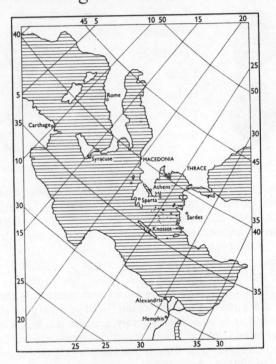

IX(1) INTRODUCTION

GREEK and Roman evidence is complementary in respect to a common standard of length. According to Pliny (ii. 21), 'a stade is equivalent to 125 paces, that is 625 feet.' The Greek stade, however, measured 600 ft. in its own scale, and these two stades

were identical if the (Greek/Roman) foot length ratio was (25/24):
this is just the relationship in which they present themselves when
the measurements (made by Stuart and by Penrose) of the
Parthenon are compared with the measurement (made by
Greaves) of the Roman pes on the monument to Statilius Aper.

Further evidence, relating to the internal structure of the Greek
and Roman systems, comes from classical authors and has been so
thoroughly summarized in modern works of reference that there
is no need here to quote from original sources. In this category of
evidence is the 8-congii rating of the Roman amphora; but the
Vespasian congius is inscribed P.X, meaning 10 librae; and so the
water-weight of the amphora must be regarded as 80 librae; and
this must be regarded also as the water-weight of the cubic pes,
because Roman tradition considered them to be equal.

Proportionally, the Greek cubic foot was a capacity for 90
librae = 60 livres of water, because the cube of its length ratio
to the Roman pes is so nearly (9/8).* Elsewhere, I have confirmed this rating by an independent calculation showing that the
cube of the metric length of the Greek foot would hold 60 livres of
water at a density of 1 kgm./litre. The potential significance
of 60 livres as the prototype talent and original unit of mass
in ancient metrology, makes the Greek cubic foot itself a unit of
exceptional interest.

IX(2) THE HECATOMPEDON

About 500 B.C. the Athenians built on the Acropolis a temple-
treasury that they called the 'Hecatompedon,' in reference,
perhaps, to the 102 Greek ft. of its interior length; later, the area of
this sacred site was doubled by the addition of a peristyle. During
the Persian wars, the whole building was destroyed, and it was
never restored. Near by, Pericles erected the new Parthenon on a
grander scale: its platform was twice the external length of the
original Hecatompedon, and exactly 100 Greek ft. in width. Its
sanctuary was larger in area than the whole of the old building in

---

* (Greek/Roman) c. ft. volume ratio $= (25/24)^3 = 1 \cdot 13028$
$$(9/8) = 1 \cdot 125$$

E.N.A.

The Parthenon, from the north-west.   The temple was dedicated in
438 B.C.   The platform supporting the columns is 100 Greek ft. in
width, and 1 myriad sq. cubits in area.

its original form; for its floor measured 6000 Greek sq. ft., and
thereon stood the great statue of Athena that Phidias sculptured
in ivory and gold.

It is on the evidence of an inscription,* published in 1890, that
modern scholarship has assigned the name Hecatompedon to the
Old Athena Temple whose foundation walls were discovered in
1886: former generations, therefore, had no hesitation in following
Plutarch, who used the words Hecatompedon and Parthenon in
conjunction when writing his life of Pericles.   Indeed, it is quite
likely that the old name attached itself to the new building.

Stuart, assisted by Revett (c. 1750), was the first to measure the
Parthenon professionally;† a century later (1888) Penrose ‡

----

* The inscription, reconstructed from 42 fragments, was first published by
H. G. Lolling in Δελτίον (1890), p. 92.   Its attribution to the Old Temple did
not pass unchallenged; in fact, it added fuel to the fire of controversy
ignited by the discovery of the temple itself.   For a summary of the
different points of view, see Appendix in *The Acropolis of Athens*, by Martin
L. D'Ooge (1908).

† *Antiquities of Athens*, by James Stuart and Nicholas Revett.   The
relevant chapter in vol. ii (1787) is entitled 'On the Temple of Minerva
called Parthenon and Hecatompedon.'

‡ *Principles of Athenian Architecture*, by F. C. Penrose (1888).

published independent measurements made during his investigation into the use of curved lines in Athenian architecture.   This evidence, in respect to the crucial hundred-foot width of the platform, is as follows:

Platform measurements in English feet; proportional Attic foot in English inches

	Platform		*Attic foot*	
	(width)	*(length)*	*(width)*	*(length)*
Stuart	101·141	227·587	12·137	12·138
Penrose	101·341	228·141	12·16	12·167

Mean of the four averages $= 12 \cdot 15$ in.
$= (81/80)$ ft.
$= (4/10k) = (1/3 \cdot 24)$ metre
$\therefore 3 \cdot 24$ Attic feet $= 1$ metre

This is the length of the Greek foot that I use throughout this analysis, because it is exactly in $(25/24)$ length ratio to the Roman

E.N.A.

Interior of the Parthenon, from the east. The sanctuary, containing Athena's statue by Phidias, had a floor area of 6000 Greek sq. ft.

foot (measured by Greaves) on the monument to Statilius Aper in the Vatican gardens: a reference by Pliny (ii. 21), to the length of the Roman stade, implies this ratio of the Greek to the Roman foot.

North, translating Plutarch in 1589, interpreted hecatompedon as meaning a square of 100 ft. side:

> For the temple of Pallas, which is called Parthenon (as a man would saye, the temple of the virgine; and is surnamed Hecatompedon for that it is a hundred foote every waye) was built by Ictinus and Callicrates.

Again, in the *Oxford English Dictionary* there is this definition:

> Hecatomped.   Measuring a hundred feet in length and breadth: a hundred foot square.   So Hecatompedon, a temple of these dimensions, as the Parthenon at Athens.

The Parthenon ground plan is not square and, therefore, lends no self-evident support to this interpretation of the word hecatompedon; but there is a more subtle way in which it may reflect ancient tradition: the length of the Parthenon platform is 2·25 times its width and, therefore, the area of the platform is 1 myriad (= 100²) Greek sq. cubits.   This interpretation of the design has not previously been noticed; but it is evident that Penrose was looking for something of the kind, for he refers to a passage where Euripides (*Ion* 1135) makes the banqueting hall 100 × 100 ft.: and he mentions the suggestion that the sacred tent at Delphi may have been 1 myriad Greek sq. ft.   There is no doubt that the inclusion of areas having significant numerical values was an objective in ancient architecture: the measurements of the Old Athena Temple, for example, seem to express a theme on the number 7.

The temple itself was divided by 7 walls into 8 chambers, and the floor of each chamber (also the ground-plan of each wall) was a multiple of 7 Greek sq. ft.   Similarly, the peristyle wall was 7 Greek ft. in thickness, and multiples of 7 Greek ft. in each exterior measurement: its interior width was 7 × 8 Greek ft., and the area enclosed by it was (7 × 8)² Greek sq. cubits.*

---

* See XIV(2) for measurements in greater detail.

IX(3) THE ROMAN FOOT
(on Statilius Aper's monument)

While Greaves was Gresham Professor of Geometry, he decided that he had 'received small satisfaction from what learned men have observed out of ancient monuments concerning the Roman foot' and must himself go to Rome. This is from his account of the journey:

> I proposed to myself in my travels abroad three ways, which no reasonable man but must approve of. And those were, first, to examine as many ancient measures and monuments in Italy and other parts as it was possible; and secondly, to compare these with as many standards and originals as I could procure the sight of, and last of all, to transmit both these and them to posterity, I exactly measured some of the most lasting monuments of the ancients.
>
> To this purpose, in the year 1639 I went into Italy, to view, as the other antiquities, so especially those of weights and measures: and to take them with as much exactness as was possible, I carried instruments with me made by the best artizans.
>
> Where my first enquiry was made after that monument of T. Statilius Vol. Aper in the Vatican gardens, from whence Philander took the dimensions of the Roman foot, as others have since borrowed it from him. In copying this upon an English foot in brass, divided into 2000 parts, I spent at least two hours. . . . Howsoever, it contains 1944 such parts as the English foot contains 2000.*

$$\therefore \text{Roman foot} = 0.972 \text{ ft.} = (3/4)k \text{ ft.}$$
$$= 11.664 \text{ in.} = 9k \text{ in.}$$

Statilius Aper died at the beginning of his career as an architect: on this monument, erected by his parents to the memory of their 'excellent son,' he is described as 'mensor aedificiorum'; and his age is given precisely as 22 years 8 months 15 days. In words of intended double meaning, a quatrain of hexameters expresses a sentiment akin to that in this paraphrase:

> Behold the harmless hog; 'twas not the Maiden's ire that laid

---

* From the *Miscellaneous Works of John Greaves*; published by Thomas Birch in 1737.

Statilius Aper's monument.
Photo. Musei Comunali. Rome.

him low, nor gutting by the fierce Mieface's sword; 'twas silent
death in swift pursuit of youth; he stole the half-built plan and
made—a ruin.

Aper's effigy stands in an alcove, and recumbent at its feet is
the dead boar (aper) to which the opening words of the quatrain
ostensibly allude.   His wife, Orcivia Anthis, is seen in an upper
panel.   Proculus, the father, was a home guard (*accensus velatus*).
The inscription dedicates this cinerarium as a vault for the ashes
of the whole family, their freedmen, and all their posterity.
The monument is supposed to have been built in the third quarter
of the first century A.D., and it is noted for its size and elaboration.

In 1743, it was presented by Pope Benedict XIV * to the
Museo Capitolino.   The linear † scale that was measured by
Greaves is on the left side of the structure.†

IX(4) Linear Standards in the Capitol

Greaves found another version of the foot in Rome.   Having
been unsuccessful in his inquiry 'after that porphyry column
mentioned by Marlianus, as also by Philander and others, with
this inscription *ΠΟΔ θ*,' ‡ he says:

My next search was for the foot on the monument of Cossutius,§ in hortis Colotianis . . . being termed by writers pes
Colotianus.   This foot I took with great care . . . afterwards
collating it with that which Lucas Paetus caused to be engraved
in the capitol in a white marble stone, I found them exactly to
agree; and therefore I did wonder why he should condemn this
with his pen (for he makes some objections against it) which

---

* In the Ashmolean Museum four marble busts by Joseph Claus (all
signed and dated 1754) commemorate Innocent XIII (1721–24); Benedict
XIII (1724–30); Clement XII (1730–40); and Benedict XIV (1740–58).

† This, unfortunately, happens to be so situated that its photograph is
not sufficiently distinct for publication.

‡ 9 ft.   The Greeks used two systems of numerical notation.   One of
these was based on an alphabet of 27 letters (including stigma ς = 6; koppa
ϙ = 90; sanpi ⌐ = 900) used seriatim in three groups of nine; for the units,
tens, and hundreds respectively.   The first group (with a distinguishing
mark) was repeated for the thousands; and M (μυριάς = myriad), with
an alphabetic numeral superimposed, was 'the orthodox way of writing
tens of thousands' (Heath).

§ Now also in the Municipal Collection.

notwithstanding he hath erected with his own hands, as appears from the inscription. . . . Now this of Cossutius is 1934 such parts as the English foot contains 2000.

Pes Colotianus (or Cossutianus) = 0·967 ft. = 11·604 in.

The presence of this foot on a tablet in the Capitol is sufficient evidence of its actual use: nevertheless, the independent evidence of the length of the Greek foot points to the pes Statilianus as the true Roman standard.

Writing a hundred years later, Martin Folkes gave this description * of the above-mentioned tablet:

> In the wall of the Capitol is a fair stone of white marble of the length of 8 ft. 5 in. English and of the breadth of 1 ft. 9 in. and a half; upon which are inscribed the standards of several measures with their respective inscriptions.

These are the measurements (converted from feet to inches) that Folkes reported to the Royal Society:

Roman foot	= 11·59 in.	Braccio di merc.	= 33·45 in.
Greek ,,	= 12·07 ,,	Staiolo	= 50·54 ,,
Braccio di tela	= 25·04 ,,	Architect's cana	= 87·9 ,,

Evidently these ratios were intended:

(Greek/Roman) foot length ratio = (25/24)
Braccio di tela : Braccio di merc. : Staiolo : : 3 : 4 : 6

These relationships are apparent:

Braccio di tela (25·04 in.) = cubit A (25·06 in.)
= 2·16 × 11·6 in.
Staiolo (50·54 in.) = 2 × 25·27 in.

The presence of the Braccio di tela raises the question whether the Cossutian foot originated as a sexagesimal submultiple of cubit A: if so then its length ratio to the Greek foot of the Parthenon was $(3/\pi)$. Writing (25/8) for π changes this ratio to (24/25); which is that of the Statilian to the Greek foot. Evidently the ratio (25/24) was applied to the Cossutian foot in order to derive the Greek foot of the tablet.

---

* *Phil. Trans.*, No. 442, p. 262 (1736).

IX(5) The Vespasian Congius

In the British Museum there is a copy of a bronze congius that was made in A.D. 75, in the reign of Vespasian; its inscription reads:

COS

IMP. CAESARE VESPAS. VI T. CAES. AUG. F. IIII

MENSURAE EXACTAE IN CAPITOLIO

P. X.

The rating P. X. means that the water-weight was intended to be 10 librae, which implies a nominal volume of 200 c. in. at a hypothetical density of $(1/20)$ libra $(= 0\cdot036$ lb.$)$/c. in. A volume of $198\cdot359$ c. in. would correspond to an amphora $(= 8$ congii) equivalent to the cube of the Roman foot rated $0\cdot972$ ft. As ancient writers attest the intended equality of the amphora and the cubic pes, Greaves (during his visit to Rome in 1639) decided to measure the cube of the pes Colotianus $(= 0\cdot967^3$ c. ft.$)$ in terms of this certified congius which was then one of the treasures in the Farnese Palace. This is his account of the experiment:

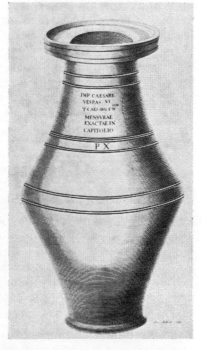

Having procured by special favour the congius

The illustration is from Greaves's drawing (1639). The congius is said to have been taken to Dresden before 1721, and it is mentioned in the Museum Guide of 1881, but I am told it is not there now. Nor is it at Naples, which has much of the Farnese collection.

of Vespasian I took the measure of it with milium (being next to water, very proper for such work) carefully prepared and cleaned. Which being done with much diligence, I caused a cube to be made answerable to the true dimension of the pes Colotianus; filling up the capacity of which, and often reiterating the same experiment, I found continually the excess of about half a congius to remain.

Allowing for the shorter pes and for some probable error in its cube, this ratio of 7·5 certainly suggests an oversize congius; and this inference is confirmed by Auzout who published * in 1680 a water-weight of 62760 French grains corresponding to (62760/6144) = 10·21 librae. Neither report mentions whether the volume of the saucer surrounding the orifice was excluded or included.

IX(6) SUMMARY OF GREEK AND ROMAN METROLOGY

Roman digit = Egyptian digit = (1/54) metre = 0·729 in.
 100000 digits = 10 stades = 6075 ft.
 = geodetic mile

„ foot = 16 digits = 11·664 in. = 0·729 ft.
 = (24/25) Greek ft.

„ stade = 625 pedes = 600 Greek ft. = Greek stade

„ mile = 8 stades = 5000 pedes = 4800 Greek ft.

Greek foot = (4/10k) metre = 12·15 in. = (81/80) ft.

„ finger = 0·76 in.; proportional ft. = 12·16 in.; proportional geodetic mile of 10 stades = 6080 ft. = Admiralty sea mile
 $0·76 = (\sqrt[3]{3})^{-1}$; $0·76 × 1·316 = 1·000$

∴„ „ × Indus inch of 1·316 in. = 1 sq. in.

„ Aeginetan iron spit = 4 Roman ft. in length
 = French aune

---

* *Mém. Acad. Roy. des Sciences*, vol. vii, part 1, p..317 (1680). (Hussey, p. 127.)

Roman jugerum = 28800 sq. pedes = 100 English sq. poles
                       = 96^2 royal sq. cubits
                       = 100^2 Sumerian sq. cubits
                       = 72 Assyrian sar = Indian biga.

Roman libra   = 12 unciae = 0·72 lb. = 5040 gt.
  ,,   uncia   = 420 gt. = 0·06 lb.
  ,,   scruple = (1/24) uncia. Siliqua = (1/6) scruple

Greek standards of mass derived from the currency:

Aeginetan iron spit = 15 unciae; drachma (iron) = 5 livres
    ,,   silver didrachm = (1/40) livre = 189 gt.
Attic (Solon)   ,,   ,,   = (5/16) uncia = (3/10) oz.
  ,,   ,,   ,, mina   = 100 drachms = 15 oz.

Greek c. ft.;   water-weight = 60 livres = 64·8 lb.
Roman amphora;   ,,   ,,   = 80 librae = 70 troy lb.
  ,,   congius;   ,,   ,,   = 10 ,,   = 7·2 lb.
  ,,   sextarius; ,,   ,,   = 20 unciae = 1·2 lb.

    Note should be taken of the 20-uncia water-weight of the sextarius in case future research reveals any evidence of the use of a 10-uncia standard in ancient metrology.

Roman c. ft.	*English c. ft.*	*Greek c. ft.*	
$(16/54)^3$	—	$(1/3·24)^3$	c. metre
= 26·02	28·317	29·403	litres
11	12	—	volume ratio approx.
8	—	9	,, ,, ,,

A Greek inscription on a stone found at Thasos reads:

The measure of the vessels 55 : 60

44 : 22

These numbers are expressed by Herodianic signs, and a published interpretation * suggests that they are interior linear

---

* By Mabel Lang in the *Bulletin de Correspondance Hellénique* (1952), part 1.

measurements (in fingers) of the height (55 : 60), maximum (44) and minimum (22) diameters of the large jar called pithos. Calculated by the formula for a conical frustum, a rough approximation to the volume for a height of 55 fingers would be:

$$\text{Volume} = (H/3)(\pi/4)(D^2 + d^2 + Dd) = (10/3)11^4$$
$$= 48803 \text{ c. fingers}$$

This suggests $12 \times 16^3 = 49152$ c. fingers = 12 c. ft. in terms of the local foot as the possible intended volume, and the published interpretation is 12 c. ft. in terms of a foot of 296 mm. called Ionic; with the 60 fingers as a permissible upper limit to the height. But 296 mm. is the length of the Roman foot, and if the 60 is interpreted as the height of a second standard its intended volume could have been 12 English c. ft. as the (11/12) volume ratio in the above table shows.

ROMAN METROLOGY

Length	Digit	Inch	Palm	Foot		Palmipes = 20 digits
	16 =	12 =	4 =	1		Cubit = 24 ,,

Foot (Pes)	Step (Gradus)	Pace (Passus)	Pole (Decempeda)	Mile (Mille passus)
5000 =	2000 =	1000 =	500 =	1

Actus = 120 pedes.   Stade = 625 pedes = (1/8) Roman mile

Area	Jugerum	Heredium	Centuria	Saltus	
	800 =	400 =	4 =	1	= 500 acres

Jugerum = 2 Actus quadrati

Mass   *Multiples of the uncia and fractions of the libra*

Uncia		Libra			Uncia		Libra	
1	=	(1/12)	=	uncia	7	=	—	= septunx
(3/2)	=	(1/8)	=	sescunx	8	=	(2/3)	= bes
2	=	(1/6)	=	sextans	9	=	(3/4)	= dodrans
3	=	(1/4)	=	quadrans	10	=	(5/6)	= dextans
4	=	(1/3)	=	triens	11	=	—	= deunx
5	=	—	=	quincunx	12	=	1	= libra
6	=	(1/2)	=	semis				

Fractions of the uncia

(1/2)	=	semuncia	(1/12)	=	semi-sextula
(1/3)	=	duella	(1/24)	=	scripulum
(1/4)	=	siculum	(1/48)	=	obolus
(1/5)	=	miliarium	(1/144)	=	siliqua
(1/6)	=	solidus			

Capacity

Sextarius		Congius		Modius		Urna		Amphora
48	=	8	=	3	=	3	=	1

Ligula		Cyanthus		Aceta-bulum		Quartarius		Hemina		Sextarius
48	=	12	=	8	=	4	=	2	=	1

GREEK NAMES OF MEASURES

Measures of length

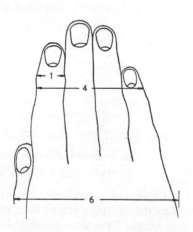

1 =	Finger	(δάκτυλος)
2 =	Knuckle	(κόνδυλος)
4 =	Palm	(παλαιστή)
8 =	Lick	(λιχάς)
10 =	Handlength	(ὀρθοδῶρον)
12 =	Span	(σπιθαμή)
16 =	Foot	(ποῦς)
18 =	Pygme	(πυγμή)
20 =	Pygon	(πυγών)
24 =	Cubit	(πῆχυς)
40 =	Step	(βῆμα)
72 =	Xylon	(ξύλον)
96 =	Fathom	(ὀργυιά)
160 =	Pole	(ἄκαινα; κάλαμος)

Also (in Greek feet):

60 =	Cable	(ἅμμα)
100 =	Plethron	(πλέθρον)
600 =	Stade	(στάδιον)
2400 =	Ride	(ἱππικόν)

Widths of the hand in terms of the width of the finger are shown on the above outline. The handbreadth calculated in VII(3) is roughly 5 fingers.

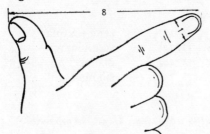

The little span that the Greeks called the lick. An allusion to table manners in the days before forks?

Measures of Capacity

Amphora (ἀμφορεύς). A large jar, with a pair of handles.

Chous (χοῦς). Originally a jug of indefinite size but later associated with the Roman congius.

Sexte (ξέστης). Corrupted from the Latin sextarius.

Subdivisions of the Sexte:

Cyathus	Oxybaphon	Cotyle	Sexte
(κύαθος)	(ὀξύβαφον)	(κοτύλη)	(ξέστης)
12	= 8	= 2	= 1

Subdivisions of the Cyathus:

Cochlear	Cheme	Mystrum	Cyathus
(κοχλιάριον)	(χημή)	(μύστρον)	(κύαθος)
24	= 12	= 4	= 1

Dry measures:

Medimnos (μέδιμνος) = 6 Hektos (ἑκτεύς)

Choenix (χοίνιξ). A slave's daily ration.

Estimated capacities of two standard measures (reconstructed from fragments) found by the American School of Classical Studies at Athens have been published (*Hesperia*, 1935, p. 347; 1938, p. 222) as 1934 and 3200 c.c. Roughly, the larger volume suggests (1/8) Roman c. ft. = congius, and the ratio of the smaller to the larger volume suggests (3/5) which would imply (1/15) Greek c. ft. with a water-weight of 6 librae. This measure is stamped on its side with the seal of the city of Athens.

One of the Arundel marbles in the Ashmolean Museum.   The monument
is of Greek provenance, but the footprint (above the man's right arm)
corresponds to the Roman foot.

IX(7) ROMAN FOOT ON A GREEK MONUMENT

Among the Arundel marbles in the Ashmolean Museum there is
one of obvious metrological import: in low relief it shows a man
with outstretched arms (the fathom), and also the imprint of a
human foot.   The intended length ratio of this fathom to this foot
appears to be 7; and the length of the foot suggests the Roman
standard of 16 digits.   The metrology may have been:

$$\text{Fathom} = 7 \times 16 = 112 = 4 \times 28 \text{ digits}$$
$$= 4 \text{ Egyptian (nominal) royal cubits}$$

This hypothetical intended length of 112 digits can also be
identified with 4 royal cubits of $20\sqrt{2}$ digits, by writing (7/5) for
$\sqrt{2}$.   That a 28-digit version of the royal cubit was in use seems
certain from the further evidence in VI(2, 3, 6, and 8).   Also, in
the Rhind Papyrus, it is used as a rating, but perhaps only for
convenience.   See VI(7).

Herodotus said the Egyptian cubit was 'the same length as the
Samian': could this monument have come from Samos?   Its
provenance is supposed to be the west coast of Asia Minor; and its
date c. 450 B.C.

Another Roman foot of Greek provenance is to be seen in the
length of the Aeginetan iron spits, when rated 4 Roman ft.

IX(8) ROMAN CENTURIATION SYSTEM

Air photography sometimes reveals alignments of modern boundaries (perpetuating Roman boundaries) that are so interrupted as to be quite invisible at ground level. In the Istrian Peninsula, such photographs show centuriation over an area measuring 10 × 4 miles round Pola: the units were squares of 20 actus side, and 200 jugera in area. North-east of Padua, squares of 15 actus side are also visible.

$$
\begin{aligned}
\text{Roman actus} &= 120 \text{ Roman ft.} = 80 \text{ Roman cubits} \\
\therefore \text{Square ,,} &= 14400 \text{ ,, sq. ft.} = 6400 \text{ ,, sq. cubits} \\
\text{Jugerum} &= 2 \text{ sq. actus}
\end{aligned}
$$

IX(9) VITRUVIUS ON MEASURING A JOURNEY

In the ninth chapter of the last of his *Ten Books on Architecture*, the author describes a Roman version of the hodometer.

The wheels of the carriage are to be 4 ft. in diameter and on one wheel a point is to be marked. When the wheel begins to move forward from this point and to revolve on the road surface it will have completed a distance of 12 + (1/2) feet on arriving at the point from which it began its revolution.

Note the use of 3 + (1/8) for π in this calculation.

Vitruvius then describes a mechanism, controlled by cog wheels, designed to release a stone audibly into a box at each completed mile; that is after each 400 revolutions of the carriage wheel moving along the road. Counting the stones gave the length of the journey.

IX(10) SOME ANCIENT COPPER INGOTS

Seventeen ancient copper ingots were found in the sea off the east coast of Euboea, near Cyme; their published * masses in grams, arranged in ascending order, are shown in column A of the

---

* J. N. Svoronos, *Journal International d'Archéologie Numismatique*, vol. ix, 1906.

following table; and against each, in column B, is the number of times it contains 315 grams, or gold-minas. The eighteenth ingot, in this list, was found at Mycenae in distant Argolis: it also is now in the museum at Athens. Numbers in brackets indicate plurality of specimens of equal mass.

A	B	A	B	A	B
5350	17	11650	37	13860(2)	44
6930	22	11970	38	17000(2)	54
9450	30	12600	40	17640	56
10080	32	12900	41	23620	75
11340	36	13230(3)	42		

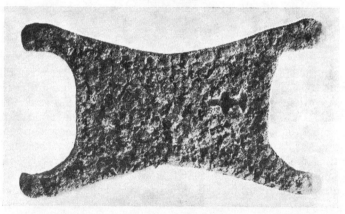

Copper ingot found at Mycenae. Note the protruding corners.
Mass 75 Gm.

All except one of the even multiples are exact, and the exception (54) is only 10 grams light; as the weights are recorded only to the nearest 10 grams, the odd multiples must also be regarded as accurate, for they lack only 5 grams each. Surely so many coincidences over such a wide range cannot have been due to chance? These ingots must have been cast from predetermined quantities most carefully measured in terms of a unit = 315 gm. = (9/10) tower lb. = (25/36) lb. = mass of a c. in. of gold.

Any required mass could have been supplied from stock if

ingots of this kind were stored in complete numerical sequence, and perhaps this was the purpose of the system.

All these ingots are slabs measuring from 1 to 2·5 in. in thickness. In plan their edges are concave and some, like that from Mycenae, have the corners extended to form handles. As contemporary wall-paintings (notably those in the tomb of Rekh-mire, at Thebes) show slaves carrying ingots edgewise on their shoulders, there can be no doubt that the concave form and protruding corners were intended to promote safe porterage. The exaggerated form of some, however, gave rise to the theory that they were designed to imitate an outstretched ox-hide, and intended to symbolize the monetary significance of the ingots by association with the more generally understood pecuniary value of the ox.

IX(11) MINOAN WEIGHTS

In the Minoan Room at the Ashmolean Museum there is a full-size copy of a remarkable weight found by Evans during his excavation of the Palace of Minos,* at Knossos in Crete. The original is 16·5 in. high, is made of purple gypsum, and is decorated in relief, on each of two sides, with the form of an octopus that stretches its tentacles over the remaining surface. There is no rating mark; but its mass has been published as 'exactly 29000 grams,' and it is certain that this would have been recorded as 64 lb. if averdepois instead of metric weights had been used.

Egyptian traders might have rated this weight as 320 debens, in terms of a deben weighing (1/5) lb.; or as 32 seps, in terms of a 2-lb. sep. Babylonian traders, however, might have regarded it as a talent of 60 minas, in terms of a mina of (16/15) lb. There is a 5-mina duck weight, in the Ashmolean Museum, that suggests the use of such a mina.

Among the smaller weights found at Knossos are two in black steatite, and similar to each other in design. The lighter (327·02 gm.) is marked with 5 small circles, meaning 5 × (1/5) librae = 326·6 gm.; the heavier (1567·47 gm.) with 2 large and 4 small circles, meaning (2 × 10) + 4 = 24 × (1/5) librae = 1567·6 gm.

---

* *The Palace of Minos at Knossos*, by Sir Arthur Evans.

A much smaller haematite weight of 12·6 gm. is exactly (1/25) Gm. = (1/36) lb.: a gold seal of 12·25 gm. is exactly (1/40) livre, rated 490 gm.; and a gold bar of 22·66 gm. is within 1 part in 1000 of being (1/20) lb. = (1/4) deben of (1/5) lb.

Ashmolean Museum copy of the octopus weight found in the Palace of Minos at Knossos. Mass =64 lb. Height= 50 Sumerian shusi = 16·5 in.

X

Pre-metric Metrology in France

PRE-METRIC metrology in France reflects aspects of the closing phase of Roman metrology in Gaul: for example, there was a land-unit called arpent that can be equated to the Roman heredium of 2 jugera; it measured 100 French square poles in terms of a pole equal to $\sqrt{2}$ English pole, and the rating of this French pole was 6 aunes. Thus, the length of the aune (ell) was equal to 4 Roman ft. and this was the length of the iron spit in the currency of Aegina before the introduction of silver coinage. Another rating for the French pole was 22 pieds, which makes the length of this pied equal to $(12/11)$ pedes; but at the time of the introduction of the metric system the official pied was about 1·5 mm. longer, and its length ratio to the Greek foot was $(100/95)$ with an error of less than 1 part in 20000.

These are the principal units in French pre-metric metrology:

Length   *Lignes*   *Pouces*   *Pieds*   *Toise*
$$864 = 72 = 6 = 1$$
Pole = 6 aunes = 22 pieds
League = 3000 toises
Pied = 324·839 mm. = 12·7889 in.

Area   Arpent = 100 sq. poles

Capacity   *Quart*   *Pot*   *Minot*   *Velte*   *Setier*   *Quarteau*   *Muid*
$$576 = 144 = 48 = 36 = 12 = 4 = 1$$
Muid = 274 litres; legal   standard (I.C.T.)
      = 268   ,,   ; actual   ,,   ,,

Mass   Lv. = Livre poids de marc   Lc. Livre de Charlemagne
    Lv. = 2 marcs   = 16 onces   = 9216 French grains
                                      = 7560 gt.
               = 18 unciae = $(3/2)$ librae = 1·08 lb.
    Lc. = $(3/4)$ Lv. = 12 onces   = 6912 French grains
                                      = 5670 gt. = 0·81 lb.
    Once = 8 drachms = 24 scruples = $(63/64)$ troy oz.

Six pieds made a toise, and it was in terms of this unit that the world was measured in order to establish the new geodetic standard now in almost universal use.   Dividing the measured length of a meridian quadrant-arc into 10 million parts called metres, the length of this metre was found to be 443·296 lines in terms of the line of which 144 made the pied.   Reciprocally, the length of the pied was found to be 324·839 mm. and then, in due course, it passed out of use.   In England the pre-metric French units were called Paris measure, and they have a special place in the history of science because they were used by Newton and other scientists of that period.   In the *Philosophical Transactions* for the year 1742 there is an account of how 'some curious gentlemen both of the Royal Society of London and of the Royal Academy of Sciences at Paris, thinking it might be of good use for the better comparing together of experiments made in England and in France, proposed sometime since that accurate standards of the measures and weights of both nations, carefully examined and made to agree with each other, might be laid up and preserved in the Archives.'   The results showed the Paris foot to be equal to 12·785 in. (the equivalent of the later official metric length being 12·7889 in.) and the Paris 2-mark weight (the livre) to be equal to 7560 gt. = 1·08 lb.

It is relevant to mention that Greaves (*c.* 1639) had found the French foot to measure 12·816 in. and had also found two other related foreign standards:

Turkish lesser pike = 25·575 in. = 2 × 12·7875 in.
Persian arish   = 38·364 in. = 3 × 12·788 in.

Evidently, therefore, the pied had an international background and Greaves discloses a still more interesting aspect of this by reporting the length of the Turkish greater pike to be (32/31) lesser pike = 2·2 ft.   This is exactly 20 Indus inches = 40 Sumerian shusi.

French metrology was so bedevilled by variations that Necker told Louis XVI in 1778 that no attempt to achieve uniformity would produce results proportionate to the trouble involved.   A case in point was the existence of two different standards for the measure of capacity called muid: one of these is now recorded as

274 litres, the other as 268 litres.   In the scale of capacities the muid was rated 576 quarts and it is relevant to note that 576 troy pints would represent a volume of 9·6 c. ft. for a troy pint rated (1/60) c. ft.   At 1000 oz./c. ft. a volume of 9·6 c. ft. would hold 1 myriad unciae = 600 lb. = 272·16 kilograms of water and, by definition, it would hold 576 troy lb. of wheat.   Similarly, (63/64)272 = 268 litres would hold 576 livres de Charlemagne of wheat of the same density, and this probably was the origin of the 268-litre standard.   The larger 274-litre standard certainly was derived from the cube of the double-pied, and it is worth noting that the intermediate muid of 272 litres mentioned above can be regarded as the cube of 2 × 324 mm.; that is the cube of 35 digits. Both French muids may have been adaptations of this hypothetical standard.

XI

Chronological History of the Metric System

FOUR distinct aims came to a focus in the metric system:

1. Standardization of a national metrology in France.
2. Use of a natural standard of length.
3. Use of a decimal scale.
4. An international system of weights and measures.

It is the international aspect of the metric system that is its present chief claim to importance; but the need for uniform weights and measures throughout France was the essential condition that made initial action possible there. The proposal to adopt a natural standard of length endowed the scheme with a sufficiently scientific aspect to attract outside attention and some co-operation; which paved the way for the subsequent acts of practical standardization that are the real foundation of its international success.

The decimal scale, which makes the metric system so convenient for some calculations, probably was the least important of the political factors: perhaps the revolutionary spirit was the chief accelerator of legislation, but the public at large would not be hustled. The metric system became the legal metrology of France in the third year of the Republic (1795); but, seventeen years later, Napoleon found popular disregard of it to be so prevalent that he sanctioned a parallel system in which the old names and customary fractions were applied to the new units. It was not until 1837 that his decree was repealed; and the use of other than the decimal metric measures became a penal offence. By that time, a new generation had been taught at school (also by Napoleon's decree) to become familiar with the new system.

In the century before the metric system was adopted there had been several advocates of a new universal metrology based on a natural unit of length: for example:

1670. Gabriel Mouton, a scientist in holy orders at Lyons, proposed a linear scale based on a geodetic minute of arc decimally divided.

1671. Picard suggested a 'universal foot,' represented by onethird of the length of a pendulum beating seconds.

1720. Cassini proposed the adoption of a geodetic foot representing (1/6000) terrestrial minute of arc.   In 1740 he made geodetic measurements that established the length of that foot, and if it had been adopted at that time its equality with the Greek foot might have been recognized by Stuart when he measured the Parthenon soon after 1750. Cassini's measurements formed the basis of the provisional metre established in 1793, but this new decimal metrology drew a veil over the sexagesimal unit and it was not until 1812 that the geodetic appearance of the Greek linear scale was noticed by Jomard.   This reference to it, however, failed to attract general attention or further inquiry.

1735–7. La Condamine, Godin, and Bouguer measured a geodetic arc of meridian, and found the length of the seconds pendulum at the equator to be 439·15 lines. These measurements were made in Peru, with a toise that became known as the Toise of Peru.

1739–40. Lacaille and Cassini measured an arc of meridian in Europe: the objective was to measure a line extending both sides of latitude 45°, and the chosen meridian was that of Dunkirk–Barcelona. These measurements showed the length of the meridian degree at latitude 45° to be 57027 toises.   Also, the length of the seconds pendulum at Paris was established as 440·5597 lines.

1747. La Condamine proposed that the length of the equatorial seconds pendulum should be adopted as a universal standard.

1778. Necker reported to Louis XVI that he had examined the means that might be employed to render the weights and measures uniform throughout the kingdom, but doubted whether the result would be proportionate to the difficulties involved.

1790. Talleyrand (then Bishop of Autun) submitted to the National Assembly a proposal to standardize the length of the seconds pendulum at 45° latitude. His proposal, having been referred to the Committee on Agriculture and Commerce, was recommended to the king, who sanctioned action on 22nd August. The French Academy of Sciences was made responsible, and appointed a committee that included Lagrange and Laplace among its members: their first report, in October, recommended the decimal division of money, weights, and measures.

Knowing that Sir John Riggs Miller, in the House of Commons during 1789, had raised the question of weights and measures, Talleyrand wrote this private letter to him on 29th March 1790.

SIR,

I understand that you have submitted for the consideration of the British Parliament, a valuable plan for the equalization of measures: I have felt it my duty to make a like proposition to our National Assembly. It appears to me worthy of the present epocha that the two Nations should unite in their endeavour to establish an invariable measure and that they should address themselves to Nature for this important discovery.

If you and I think alike on this subject, and that you are of opinion that much general benefit may be derived from it, it is through you only that we can hope for its accomplishment; and I beg to recommend it to your consideration. Too long have Great Britain and France been at variance with each other, for empty honour or for guilty interests. It is time that two free Nations should unite their exertions for the promotion of a discovery that must be useful to mankind.

I have the honour to be, Sir, with due respect, your most humble and obedient servant,

THE BISHOP OF AUTUN.*

In the next session of Parliament Miller reported the

---

* Miller's translation.

receipt of this letter, and expressed himself in favour of the scheme; but nothing came of it. A few months later the bishop sent Miller a copy of the National Assembly's minute of 8 May, in which Louis XVI is asked to write to George III inviting joint action to determine a natural standard of weight and measure: but I am informed by the Registrar that there is no such letter in the Royal Archives at Windsor Castle.

1791. 'The King and Queen desiring to make their Easter Communion at St. Cloud were turned back by the mob.' *

Second report by the committee recommending that the standard of length should be:

One ten-millionth of the meridian quadrant.

This length was to be determined by calculations based on the measurement of a meridian arc extending from Dunkirk to Barcelona; that is an extension of the line measured by Lacaille and Cassini in 1739. The decimal aspect of the system, recommended in their first report, was emphasized by discarding the traditional degrees and minutes of angular measurement. The committee rejected the pendulum, on principle, because it involved time as a non-linear element. The Academy of Sciences adopted the committee's recommendations.

1792. French Republic established on 22 September, the day of the autumnal equinox.

Mechain and Delambre began work on the meridian measurement.

1793. Louis XVI executed, on 21 January.

Third report of the committee; recommending the name 'metre' for the new linear unit, and giving its length provisionally as 443·44 lines. This provisional length was calculated from Lacaille's measurement; which showed the length of the meridian degree at latitude 45° to be 57027 toises.

---

* *A History of Europe*, by H. A. L. Fisher, p. 805.

Meridian quadrant $= 57027 \times 90 = 5132430$ toises
$= 5132430 \times 864 = 443 \cdot 44 \times 10^7$ lines

The Academy submitted its committee's report to the Convention, which had replaced the National Assembly: this report was adopted by decree on 1 August, and a brass standard of the provisional metre was made: it is preserved in the Conservatoire des Arts et Métiers at Paris.

War with England began in this year.

1794. Execution of Robespierre, and end of the Terror.

1795. The metric system became the legal metrology of France by the Law of 7 April 1795. The franc was introduced into the coinage by this law.

1798. Mechain and Delambre having (with great difficulty, owing to the political conditions) established the difference in latitude between Dunkirk and Barcelona as 9° 39', and allowed for the flattening of the earth, calculated the length of the meridian quadrant to be 5130740 toises. This was only 1 part in 3000 less than the calculated length based on Lacaille's measurement of a degree, and reduced the length of the provisional metre by only 0·146 line to:

Final metre $= 5130740 \times 864/10^7 = 443 \cdot 296$ lines
$\therefore$ French foot $(= 144 \text{ lines}) = 0 \cdot 324839$ metre

Three platinum standards and several iron standards of the metre were now made.

The sub-committee, charged with the responsibility of making a standard of mass to represent a cubic decimetre of distilled water at the temperature of melting ice, also completed its work this year; and a platinum kilogram was constructed equal to 18827·15 French grains in the scale of the Pile de Charlemagne. In this scale the livre poids-de-marc was rated 9216 French grains: the mass of the kilogram, therefore, was equivalent to $(18827/9216) = 2 \cdot 0421$ livres, and the mass of the livre was 489·69 gm.

Battle of the Nile.

1799. The new standards were presented formally to the Corps Législatif, and legalized by statute abolishing the provisional standards. One of the three new standard metres became known as the Metre of the Archives. The Government had intended to make a distribution of copies of the new standards but found the cost too high. For those in possession of copies of the Toise of Peru, the metre was represented by (443·296/864) of its length.

1800. A decree authorizing the use of a more popular nomenclature was issued.

1812. Public prejudice against the new scheme being still very strong, a parallel system was established in which the old names and customary fractions were applied to the new units: thus, the length of 2 metres was called a toise and divided into 6 pieds that were (1/3) metre in length. At the same time, a decree enacted that the legal decimal system must be taught in schools and used in all official transactions.

Retreat from Moscow.

1837. The decree of 1812 was repealed, and it was enacted that the use of measures and weights other than those of the metric system would become a penal offence from the beginning of 1840.

1864. In England, Parliament passed an Act (27 and 28 Victoria, c. 117) legalizing the use of metric metrology in contracts; and providing a schedule of authorized equivalents: but this Act did not legalize the use of actual metric weights in trade. The authorized equivalents included:

1 metre =	39·3708 in.	1 kgm. =	15432·3487 gt.
1 are =	119·6033 sq. yds.	1 litre =	1·76077 pints

1867. A convention of the International Geodetic Association recommended the international use of the metric system in geodetic work, and advocated the construction of a new European prototype metre (differing as little as possible from the Mètre des Archives) to be available for international use, under the supervision of an international bureau.

1870. Franco-Prussian war.

1872. An International Commission, convened by the French Government, met at Paris and advocated the construction of international metric standards, to be kept by an international bureau located at Paris.

1875. The Metric Convention. Eighteen countries signed a treaty to establish and maintain an international bureau of weights and measures. England was not one of these signatories: the Warden of the Standards was a delegate to the 1872 Commission, but was not allowed to participate in subsequent events because 'Her Majesty's Government declared that they could not recommend to Parliament any expenditure connected with the metric system, which is not legalized in this country, nor in support of a permanent institution established in a foreign country for its encouragement. They have consequently declined to take part in the Convention or to contribute to the expenses of the new Metric Bureau, and have directed the Warden of the Standards to decline being appointed a member of the new International Committee or to take part in the direction of the new International Metric Bureau.' *

1884. Great Britain joined the Metric Convention.

1897. 60 and 61 Victoria, c. 46. 'An Act to legalize the use of weights and measures of the metric system.' This Act permits the use of metric weights and measures in trade, and requires the Board of Trade to include metric denominations among its standards.

---

* *On the Science of Weighing and Measuring*, by H. W. Chisholm (Warden of the Standards), 1877.

Historical Aspects of English Metrology

Winchester City Museums photo of the 8-lb. Elizabeth I averdepois weight in the Westgate Museum. The Elizabethan standards (averdepois and troy) proclaimed in December 1587 retained their legal status until 1824.

XII(I) HISTORY OF THE ENGLISH STANDARD POUND

THE Weights and Measures Act 1878 is the legal basis of modern British Metrology. It incorporated the recommendations of a Commission appointed in 1838 to 'consider the steps to be taken for restoration of the standard of weight and measure' rendered necessary by loss and damage of former standards in the fire that destroyed the Houses of Parliament in 1834.* The utmost care was taken to reproduce exactly † the magnitudes established by the previous Act of 1824: but whereas under that Act a certain troy weight became the legal physical standard of reference, in the Act of 1878 a newly made averdepois weight was given that status.

This apparently drastic change was no more than legal confirmation of the exclusive commercial position that the lb. averdepois had long held in this category of mass—a position that was emphasized in 1841, when the above-mentioned Commission reported that:

> The troy pound is comparatively useless even in the few trades and professions in which troy weight is commonly used and to the great mass of the British population it is wholly unknown. The statements of medical men and those persons concerned in the trade of bullion show that even to them the troy pound is useless. The avoirdupois pound on the other hand is universally known throughout this kingdom. We feel it our duty therefore to recommend that the avoirdupois pound be adopted instead of the troy pound as the standard weight.

Actually, the use of the troy pound at that time was already of only nominal legality; for in 1834 it had been found necessary to pass an Act to 'amend and render more effectual' the Act of 1824, and therein it was declared that:

---

* This fire on 16 October was caused by excessive stoking: the House had ordered the destruction of a store of old tallies, and they were being burnt in the heating stoves. Tallies were made by splitting lengthwise sticks that had been notched in code to record sums of money; one piece was given as a receipt, the other was retained in the Exchequer.

† For an account of the scientific principles involved, see report by W. H. Miller, *Phil. Trans.* (1856), vol. cxlvi, p. 753.

All goods sold by weight shall be sold by avoirdupois weight except gold, silver, platina, diamonds and other precious stones which may be sold by troy weight: and drugs which when sold by retail may be sold by Apothecaries' weight. (4W. IV. c. 49. xiii.)

It may well be asked, therefore, how the troy pound came to be adopted in the Act of 1824; and the answer is that the actual weight of reference * had been in existence already for 66 years. It was made in 1758, and would have been legalized in 1760 if the relevant bill had not been interrupted in mid passage by the general disturbance of parliamentary business caused by the sudden death of George II.   An abortive attempt to reopen the subject was made in 1789; and again in 1790 when the same member endeavoured to arouse interest in the state of the English standards by reading a private letter from Talleyrand about the metrological situation then developing in France.   In 1814 another advocate drew the attention of the House to the fact that the 17th article of the Act of Union (1707), requiring the Scottish weights and measures to be the same as the English, could not be carried into effect because of the 'uncertainty of the English standard.'   A supporter of the motion to appoint a committee remarked, somewhat cynically, that he hoped the hon. member 'would be more successful in the attainment of his purpose than his predecessors had been in the same effort'; and in fact the subsequent Bill did pass the Commons, but it was thrown out by the Lords in 1818.   Nevertheless, there ensued in that year the appointment of the Commission whose recommendations, after scrutiny by a Select Committee in 1821, were embodied in the Act of 1824.   This committee remarked that they

concur with the Commissioners on the inexpediency of changing any standard which already exists in a state of acknowledged accuracy: also concur in recommending that the subdivisions employed in this country be retained, as being far better adapted to common practical purpose than the decimal scale.

The troy standard weight of 1758 was made for a committee appointed in that year to 'inquire into the original standards of

---

* The mass of the legal weight of reference was 2 lb. troy.

weights and measures in this kingdom, and to consider the laws relating thereto.' At that time, the standard weights (in the Chamberlain's office of His Majesty's Exchequer) were still those made under the direction of a jury appointed in 1582, and legalized by royal proclamation dated 16 December 1587. These Elizabethan weights, which remained the legal English standards for 237 years (1587–1824), came into existence as the result of complaints

> that the weights being used throughout this realm are uncertain and varying one from another to the great slander of the same Our realm and decency of many, both buyers and sellers.

| 91 lb. | 56 lb. | 28 lb. | 14 lb. |

Winchester City Museums photo of the Edward III averdepois weights in the Westgate Museum. The heaviest is the quarter-sack of 91 lb.; the lightest of those shown above is the 14-lb. stone, but the complete set includes a pair of 7-lb. cloves. See p. 25.

A jury was appointed to conduct an inquisition; but its report of 1574 confused the pound averdepois of 7000 gt. with the libra mercatoria of 15 oz. troy (7200 gt.), and the new standards that it sponsored were so soon discredited that a new jury was empanelled in 1582: it proceeded at once to make the

> troy standard to agree with the true proportion of the pile or standard remaining in the Goldsmiths' Hall to size all other

weights by which hath been of ancient continuance and use, and
the averdepois with the true proportions of an ancient standard
of 56 lb. remaining in the Exchequer, as it seemed, since the time
of King Edward III, to size all other weights by.

Thus the Goldsmiths' pile, which the previous jury had
reported to be 'of no credit,' was confirmed in its traditional
importance; and the 4-st. weight of Edward III was adopted as
the basis of new averdepois standards: these were the weights,
legalized by Elizabeth's proclamation of 1587, that remained the
sole legal standards of reference until the Act of 1824.

Summarizing: these are the principal aspects and phases of the
history of the standard pound.

Winchester City Museums photo of three Elizabeth I nesting troy weights
in the Westgate Museum; they are dated 1588. The complete set forms a
dimidiated series from 256 troy oz. to 1 troy oz. The troy system was
based on the ounce and this was its usual form: it included the 8-oz. mark,
but the pound was represented only by the 8 + 4-oz. combination.

Long-established tradition was focused in the Edward III
averdepois weights at the Exchequer and in the pile of troy
weights at Goldsmiths' Hall, but variation in commercial usage
led to the Elizabethan inquisition. The first jury reported in
1574 and the second jury was appointed in 1582; their new standards (averdepois and troy) were legalized in 1587 and they were
not superseded legally until the Act of 1824.

From the reign of Henry VIII, troy weights had been in use at the mint (also from earlier times in a wider field) and this association with the coinage gave the system pre-eminent status. A new troy weight was made in 1758 for a Parliamentary Committee, but the legalizing Bill was interrupted in 1760 by the death of George II: it was this troy weight of 1758 that was legalized by the Act of 1824. The legal use of the troy system was limited by the Act of 1834 and the standards were destroyed in that year by the fire that destroyed the Houses of Parliament.

Averdepois weights had been used in commerce from time immemorial; the Act of 1834 confirmed this status and the Commission appointed in 1838 recommended in 1841 that the averdepois pound be adopted as the legal standard of reference. A new weight (made with a mass of 7000 gt. in terms of the grain in which the troy pound was rated 5760 gt.) was legalized by the Act of 1878.

Historically, the averdepois system in England is first apparent in association with the stone of 14 lb. but elsewhere it goes back to the oldest weight in the world—the Babylonian mina D of 1·5 lb. This ancient weight also emphasizes the importance of the ounce and its relation to the uncia.

XII(2) STEELYARD AND BALANCE

There were two methods of weighing: by the steelyard, using auncel weights; and by the balance, using true weights. In the former, the longer arm is graduated like a yard-stick or ell (French, aune), and false weighing was not only possible but prevalent: so much was the system abused that Edward III (in 1351) made it illegal. A reference to false weighing occurs in William Langland's *Vision of William concerning Piers the Plowman*, which is supposed to have been written between 1362 and 1399. A. Passus V (lines 117–18; 131–2):

Furst I leornede to lyze • a lessun or tweyne
And wikkedliche for to weie • was myn other lessun

. . . . . .

The pound that heo peysede by • peisede a quartrun more
Then myn auncel dude • whon I weyede treuthe.

XII(3) Averdepois

The following is an example of the use of the word averdepois:

1353. 27 Edward III, c. 10. Because we have perceived that some merchants do buy averdepois wools and other merchandises and sell by another . . . we do will and establish . . . that wools and all manner averdepois be weighed by the balance, so that the toung of the balance be even.

A guild of importers of pepper (gilda piperariorum) is said to have been formed about 1180; and its members, with those of other gilds concerned with similar trading customs, became known as men of averdepois. About 1365, however, they began to be called grocers, and in due course their company (mestere grossariorum) acquired its traditional responsibilities in respect to the King's weigh-house and beams within the city of London.*

XII(4) The Tower Pound

Prior to 1527 the pound in use at the mint at the Tower of London was a pound of 5400 gt. known as the tower pound; and it was in terms of this pound that the king's indentures † specified the number of coins of each denomination to be shorn from the pound of bullion. Henry VIII abolished the use of this pound, in favour of the troy pound of 5760 gt.; the relevant Exchequer minute of 30 October 1527 reads:

And whereas heretofore the merchants paid for coinage of every pounde Towre of fyne gold, weighing XI oz quarter Troye ii s vi d. Now it is determined by the King's Highness and the said councelle, that the foresaid pounde Towre shall be no more used and occupied, but al maner of golde and sylver shall be wayed by the pounde Troye, which maketh XII oz Troye, which exceedeth the pounde Towre in weight iii quarters of the oz.

$$\therefore \text{ (Tower/Troy) pound mass ratio} = (11\cdot25/12) = (15/16)$$

Probably it is the tower pound that is meant by the following reference to the pound of the mint, in De Institutis Lundonie,

---

* See XV(4): Kingdom. † See XIV(6).

promulgated during the reign of Ethelred the Unready, A.D. 968–1016.

> And those who have charge of towns shall see to it, under pain of incurring the fine for insubordination to me, that every weight is stamped according to the standard employed by my mint; and that the stamp used for each of them shall show that the pound contains 15 ores.*

$$\therefore \text{ Hypothetically, 15 ores} = 5400 \text{ gt.} = \text{tower lb.}$$
$$\therefore \qquad\qquad \text{Ore} = 360 \text{ gt.}$$

This ore provides a common rating unit linking the tower, troy, and mercantile pounds with the libra and the livre.

In one of the treaties with the Danes (made by Edward and Guthrum) there are references to fines, from which it seems likely that:

$$21 \text{ ores} = 60 \text{ shillings} = 3 \text{ half-marks}$$

This would make the ore equivalent to 5 shillings, or to 20 pence in terms of the Mercian fourpenny shilling; and the proportional pennyweight would be 18 gt. Some support for this hypothesis is apparent in the actual weight range of extant silver pennies struck by Offa of Mercia (A.D. 757–96). These were similar in general form to the new denarius, first struck in France by Pippin the Short (A.D. 752–68). See XIV (6).

The Mercian fourpenny shilling was still current (as money of account) at the time of the Conquest; evidence of this appears in one of the so-called laws of William I:

> If it happens that a man cuts off the hand or foot of another, he shall pay him half his wergeld according to his inherited rank. For the thumb he shall pay half the value of his hand; for the finger next the thumb 15 shillings according to the English reckoning, that is four pence to the shilling.

In course of time a twelvepenny shilling became customary as money of account: this rating had already taken root in France; possibly as the Carolingian silver equivalent of the Merovingian gold triens.

---

* *The Laws of the Kings of England from Edmund to Henry I* (1925). Edited and translated by Agnes Jane Robertson.

These references to monetary evidence are made here for their metrological significance, and because it was customary in France and in England to call the pound of mass by its monetary rating. The tower pound became a pound of twenty shillings, in terms of a twelvepenny shilling; and two marks tower (= 16 oz. tower) became a pound of 25 shillings troy.

XII(5) ROMAN METROLOGY IN THE STONES, FOTMAL, SACK, AND LOAD

In the *Tractatus de Ponderibus et Mensuris* (*c.* 1303) there is a curious piece of metrological arithmetic that can be summarized thus:

$$\begin{aligned}
\text{Pound} &= 25 \text{ shillings} & = &\quad \text{M} \\
\text{Stone} &= 12 \text{ lb.} & = &\quad 12 \text{ M} \\
\text{Fotmal} &= (6 \text{ st. } 2 \text{ lb.}) & = &\quad 70 \text{ M} \\
\text{Cartload} &= 30 \text{ fotmals} = (30 \times 70) \text{ M} & = &\quad 2100 \text{ M} \\
&= (8 \text{ score} + 15) \text{ st.} = 175 \text{ st.}
\end{aligned}$$

Proof: $(6 \times 30) = (9 \times 20) = 180$ st.

Subtract $(30 \times 2) = 60$ lb. $= \underline{\quad 5 \quad}$,,

$\therefore$ Cartload $= 175$,,

Interpretation:

M = Mercantile pound = 25 shillingsweight troy
    = 15 oz. troy = 16 oz. tower = 2 marks tower
$\therefore$ Stone   = 12 M = 16 tower lb. = 15 troy lb.
$\therefore$ Fotmal = 70 M = 100 librae   = 72 lb.
$\therefore$ Load   = 30 fotmals = 3000 librae = 1000 minas N

Libra	M	Stone	Fotmal	Load	
3000	2100	175	30	1	= 2160 lb.
100	70		1		72 lb.
	12	1			$\begin{cases} = 16 \text{ tower lb.} \\ = 15 \text{ troy lb.} \end{cases}$

Elsewhere, in the same document, the sack of wool is said to be equivalent in mass to one-sixth of a cartload of lead; and to weigh 28 stones in terms of a stone of 12·5 M = 200 tower oz.; not the

12-M stone of the above table. In tabulated form, the picture of the 12·5-M stone is:

Libra		M		Stone		Wey		Sack		Load		Last		
6000	=	4200	=	336	=	24	=	12	=	2	=	1	=	4000 livres
3000	=	2100	=	168	=	12	=	6	=	1			=	2160 lb.
500	=	350	=	28	=	2	=	1					=	360 lb.
250	=	175	=	14	=	1							=	2·5 fotmals
		12·5	=	1									=	200 tower oz.

A further vestige of Roman metrology appears in this passage from the translated version published in the *Statutes of the Realm*:

> A sack of wool ought to weight 28 stone, that is 350 pounds, and in some parts 30 stone, that is 375 pounds, and these are the same according to the greater and lesser pound.

Interpretation:

Sack = 28 stones of 200 tower oz. = 500 librae
= 30 ,, ,, 200 Roman unciae = ,, ,,
'Greater pound' = 16 tower oz. = tower dimark = M
'Lesser pound' = 16 Roman unciae = Roman double-bes

The Roman mark of 8 unciae was called bes. For convenience, I call 2 marks tower, a dimark.

There is yet another stone in the *Tractatus*: its mass was 25 librae, and 120 such stones made the cartload of lead; also, as the above table shows, 10 of these stones made the wey.

Finally, reference must be made to 25 Edward III c. 9 (1351) for the sack of 26 st. in terms of the stone of 14 lb. averdepois. The mass of this sack is in (91/90) mass ratio to the sack of 500 librae, and it weighs 364 lb.: sets of weights, issued to Winchester and to the other cities in this reign, included a quarter-sack weight of 91 lb.

This ratio (91/90) also expresses the mass ratio of the modern fodder to the old cartload of lead, and reflects its transition to the averdepois scale: the customary fodder of lead weighs 19·5 cwt. = 156 st. = 2184 lb. and, in terms of the 26-st. sack, it maintains its traditional rating of 6 sacks.

In addition to the quarter-sack of 6·5 st., the Winchester series of Edward III weights includes the half-hundredweight (= 4 st.), the quarter (= 2 st.), the stone, and a pair of cloves that provided a balanced load when transported by a rider on horseback. As the smallest unit in this early manifestation of the averdepois system, the clove of 7 lb. was responsible for the (91/90) sack ratio and its obvious importance in English metrology provokes curiosity about its antecedents, it may be relevant, therefore, to draw attention to the fact that it can be rated 10 Gp. = 49000 gt. Similarly, the ton can be rated 320 cloves = 3200 Gp.

XII(6) THE WIRKSWORTH DISH

Lead mining in Derbyshire was active during the Roman occupation, probably also in earlier times, and the traditional right of the individual to dig for lead in this district is associated with customs that are amusingly summarized in Manlove's *Rhymed Chronicle*: * he was steward of Wirksworth Barmote in 1653. The stewards appointed the barmasters responsible for measuring the output on which duties called lot and cope were payable; and in Wirksworth there still exists a brass dish that was made in 1512 as a standard of reference for this purpose.†

Its capacity for lead ore as mined there 'has long been established at about 65 lb.,' ‡ which suggests that its original rating may have been 64·8 lb. = 90 librae = 60 livres.

---

* *Reprinted Glossaries*. English Dialect Society.
† There is a copy of this dish in the Science Museum at South Kensington: in 1877 it was in the Museum of Practical Geology, but its earlier history is unknown.
‡ T. L. Tudor in the *Journal of the Derbyshire Archaeological Society*, 1937–8. The illustration of the dish at Wirksworth is reproduced from the Journal, by permission of the Society.

The ore was measured with wooden copies of the standard; and the practical limit to the capacity of such a dish would be the weight that the user could lift frequently without undue fatigue.

Roughly, the brass dish is 4 in. deep and 21·5 by 5 in. in interior length and breadth: its sides are sloped to facilitate moulding. An inscription, covering, both sides, reads:

> This dish was made on the fourth day of October in the fourth year of the reign of King Henry VIII—before George Earl of Shrewsbury, steward of the king's most honourable household and also steward of all the honour of Tutbery—by the assent and consent alike of all the miners as of all the brenners within and adjoining the lordship of Wirksworth, parcel of the said honour. This dish to remain in the Moot Hall at Wirksworth, hanging by a chain so that the merchants or miners may have resort to the same at all times to make the true measure after the same.

It is a curious fact that its validity as a standard of reference will cease if the dish itself is lost or destroyed; in that event the Mining Customs Act of 1852 imposes on the wapentake the 15-pint standard that was confirmed for the High Peak section of the King's Field by an Act of 1851.

The first verse of Manlove's *Rhymed Chronicle* runs thus:

> By custom old in Wirksworth Wapentake
> If any of this nation find a rake,
> Or sign, or leading to the same; may set
> In any ground, and there lead ore may get.
> They may make crosses, holes, and set their stowes,
> Sink shafts, build lodges, cottages and coes;
> But churches, houses, gardens all are free
> From this strange custom of the minery.

A record of these liberties 'with extracts from the bundles of the Exchequer ... taken in the reign of Edward I and continued ever since' * was printed and published in 1645. In the findings of a commission of 16 Edward I (1288), each miner is entitled to 4 perches of ground plus 7 ft. for his shaft.

---

* L. B. Williams, *Mining Magazine*, August 1932.

XII(7) The Winchester Bushel

Winchester's ancient status as the capital is reflected in the attachment of its name to the equally ancient (but now obsolete) standard corn bushel and proportional gallon. Extant Winchester bushel measures made in the reign of Henry VII are to be seen in the museums at Winchester, Salisbury, and Norwich. They are large bronze bowls, cast with a pair of handles and three short feet: around the outside is this inscription in raised Gothic lettering: 'Henricus Septimus dei gratia rex Hanglie et Francie.' A hound, a portcullis, and the Tudor Rose form decorative punctuations in this text; and unpremeditated diversity was achieved by the moulder who made the inscription on the Salisbury bowl read 'dei gra rextia rex,' thus revealing the use of patterns containing about three letters in a group so that they could be withdrawn readily from the mould without much risk of disturbing the sand.

The history of these bowls began in 1491 when Parliament petitioned the king to make 'weights and measures of brass

Winchester City Museums photo of the Winchester bushel measure in the Westgate Museum. It was made in the reign of Henry VII, and copies were issued to the cities and principal towns.

Moulder's error in the inscription on the Winchester bushel measure issued to Salisbury. The lettering reads 'dei gra rextia rex,' instead of 'dei gratia rex.'

according to the very true standard . . . at your p'pre cost and charge.' In 1495 they were ready and delivered to over forty cities and towns, but later someone 'upon more diligent examynacion' found them to be defective. It must have been humiliating to have had to pass 12 Henry VII, c. 5, ordering their return, but a businesslike endeavour was made to minimize further delay: this time, however, the work was to be done 'att the costis and chargis of the said cities.'

In order that there should be no doubt about the 'very true standard' while the second set of measures was being cast, the Act recited an inappropriate stock formula defining the bushel of 8 troy gall.! This was a very much smaller bushel than the one they were trying to make; in fact it was the wine bushel, also called the London bushel.

If the extant bushels represent the second attempt then it must be admitted that this also was defective, for it gave a bushel of less than 8 gall. in terms of extant examples of the gallon measure issued in the same reign and, presumably, at the same time. In 1688, the Commissioners of Excise reported to the Treasury that they had found at the Exchequer:

> 3 standard gallons: the one in the time of Henry the 7th, and two others dated 1601 in the 44th year of Queen Elizabeth, which we have caused to be tried by able artists and found them to be of equal contents and to contain 272 cubic inches.

Winchester City Museums photo of the Henry VII gallon measure in the Westgate Museum. This is the Winchester corn gallon, rated one-eighth of the Winchester bushel.

In 1695 the Parliament of Ireland legalized a gallon of 272·25 c. in. and a proportional bushel of 8 such gall. This evidence, taken in conjunction with the farming tradition that rates the mass of a bushel of wheat at 4·5 st., makes me think that the true Winchester standard was:

Winchester bushel (hypothetical original capacity):

$$= 4\text{·}5 \text{ st. } (= 63 \text{ lb.}) \text{ of wheat at } 50 \text{ lb./c. ft.}$$
$$= (63/50) = 1\text{·}26 \text{ c. ft.} = 2177\text{·}28 \text{ c. in.}$$
$$= 8 \text{ gall. of } 272\text{·}16 \text{ c. in.}$$

When expressed to the nearest quarter of a cubic inch this hypothetical gallon becomes the gallon of 272·25 c. in. legalized in 7 William III, c. 26 (Parliament of Ireland). The proportional bushel is 2178 c. in. Nevertheless, in the following year the Parliament in England declared (in an Act for laying a duty on malt) that an entirely new geometric standard should be called the Winchester bushel. The relevant clause in 8 & 9 William III, c. 22 (1696–7) reads:

It is hereby declared that every round bushel with a plain and even bottom being eighteen inches and a half wide throughout and eight inches deep shall be esteemed a legal Winchester bushel according to the standard at His Majesty's Exchequer.

The equivalent volume (not disclosed in the Act) is 2150·4 c. in. = 1·244 c. ft., and certainly this was intended to perpetuate (cylindrically, to the nearest linear half inch) the 2145·6 c. in. found by measurement to be the volume of an extant Henry VII Winchester bushel at the Exchequer. This measurement was made in the presence of certain members of Parliament who were taking a particular interest in the debate, and it is clear evidence of the discrepancy between the Henry VII bushel and gallon standards that have survived.

The cylindrical version of the Winchester bushel became and remains (on a volumetric basis) the standard bushel of the United States of America.

Of the bushel rated 1·26 c. ft. it may be interesting to note that it exceeds the cube of the Assyrian foot by less than 1 part in 400.

XII(8) THE OLD WINE GALLON

One of the most important early documents relating to English weights and measures is the *Tractatus de Ponderibus et Mensuris* (*c.* 1303): its opening paragraph (as translated in the *Statutes at Large*) reads:

> By consent of the whole realm, the King's measure was made so that an English penny, which is called Sterling, round without clipping, shall weigh thirty-two grains of wheat dry in the midst of the ear; twenty pence make an ounce and twelve ounces make a pound and eight pounds make a gallon of wine and eight gallons of wine make a bushel of London, which is the eighth part of a quarter.

The context and other relevant evidence require that this definition of the gallon be interpreted as meaning that 8 lb. of wheat measure a gallon for wine. The Latin version mentions corn, but not wine: 'Et octo librae frumenti faciunt galonem.'

At a hypothetical bulk density of 60 troy lb./c. ft. (which is within the range for wheat) the volume of this troy gallon would be 230·4 c. in. and it would hold 10 troy lb. of wine at a hypothetical density of 250 gt./c. in. It acquired a volume of 231 c. in. as a cylinder 7 in. diameter × 6 in. deep and it is still in use as the standard gallon of the U.S.A.

This volumetric rating of 231 c. in. is derived from the geometry by the use of (22/7) for π.* It has no independent validity to conflict with 230·4 c. in. as the hypothetical equivalent of the earlier wheat-weight rating, but it does imply that the cylindrical form was already current in 1688, for in that year the Commissioners of Excise were much perturbed to receive 'an intimation that His Majesty hath lately been informed that the true standard gallon contains only 224 c. in. and not 231 c. in. as hath been used.' †

Seeking to justify their measure, the Commissioners inspected

---

* Volume of cylinder $= LD^2(\pi/4) = 6 \times 7^2(22/28) = 231$.
† *Excise to Treasury Correspondence*, vol. iii. Library of H.M. Customs and Excise.

those at the Exchequer and found the gallons * to be much larger than their own. So

> We applied ourselves to the Guildhall of the City of London where we were informed the true standard of the wine gallon is and hath always remained and there we found an ancient standard wine gallon which we have caused to be tried by the said artist and find the same to contain only 224 cubic inches. We further find that according to the said standard wine gallon remaining in the said Guildhall the standard galls of all other parts of the Kingdom for wine are and have been made and granted and are used in all parts. All which we humbly lay before your Lordships and humbly crave your Lordships' directions therein.†

Such was the text of the letter in which the Commissioners of Excise informed the Treasury of their embarrassment. The Treasury referred to the Attorney General for an opinion, and his sage reply was that:

> We cannot resort to Guildhall for a standard, for no Law or Statute refers to that, or makes that the standard, nor do I know by what Law it can be imposed upon such as are not willing to receive it.

So much for the Guildhall gallon: the measures at the Exchequer, however, were in quite a different legal category, but Powys dismissed them with this politic remark:

> We must not resort to the Exchequer for a standard to which almost all the Acts of Parliament refer, for there is none there but what the King will be vastly a loser by.

Then, after commenting on the difficulty of interpreting in terms of wine the wheat-weight rating of the gallon ‡ that is recited in the medieval laws, he said:

> Though I do not know or see how 231 cubic inches came to be taken up and settled as the contents of a wine gallon, yet I do not think it safe to depart from it now usage hath settled it.

In 1700 the exclusive validity of this wine gallon was

---

* These were the Winchester gallons of 272 c. in. mentioned in XII(7).

† The wine-weight of this Guildhall gallon can be rated 8 lb., see XII(9), and, therefore, probably was an interpretation (in the averdepois scale, and ignoring any connection with wheat) of the definition in the *Tractatus*.

‡ The very gallon that justified his own decision.

challenged from the opposite side.   An importer, prosecuted for refusal to pay balance of duty required by the Excise according to their reckoning, stated in court that he had declared his consignment in terms of the Exchequer gallon * and contended that this was a customary wine gallon.   The authorities rightly contended that this gallon of 4 Exchequer quarts was a standard only for ale and beer: however, they dropped the case, having decided that the status of the Excise gallon must be strengthened by legislation.   This was not done at once, but the need became imperative when the Act of Union required † 'the same weights and measures to be used throughout the United Kingdom as are now established in England,' and the following article ‡ appeared in another Act passed that year:

> Be it further enacted . . . that any round vessel (commonly called a cylinder) having an even bottom and being seven inches diameter throughout and six inches deep from the top to the inside of the bottom or any vessel containing two hundred and thirty-one cubical inches and no more shall be deemed and taken to be a lawful wine gallon and it is hereby declared that two hundred and fifty-two gallons consisting each of two hundred and thirty-one cubic inches shall be deemed a ton of wine and that one hundred and twenty-six such gallons shall be deemed a butt or pipe of wine and that sixty-three such gallons shall be deemed an hogshead of wine.

This traditional 252-gall. rating of the ancient tun supports the hypothetical 10-troy-lb. rating for the wine-weight of the Excise gallon, for it can be interpreted as $10 \times 252 = 36 \times 70$ troy lb. $= 36 \times 80$ librae $= 36$ Roman amphorae $= 32$ Greek c. ft.

In 1824, a gallon having a water-weight of 10 lb. averdepois became the legal standard and thereafter, in England, the Old Wine Gallon became a memory with its past fame enshrined in the friendliness of its present title, but elsewhere it is not dead yet, for long ago it began a new life in the New World and there it flourishes with undiminished prestige as the standard gallon of the United States.

---

* Reckoned as 4 quarts of 70·5 c. in. = 282 c. in.   See XII(9).
† 6 Anne, c. 11, art. 17 (1706).   Union effective 1 May 1707.
‡ 6 Anne, c. 27, art. 22 (1706).   Text, *Statutes of the Realm*.

XII(9) THE STANDARD GALLON

The present standard gallon, defined in the Act of 1878 as a capacity for 10 lb. of distilled water at 62° F. when weighed under specified conditions, was legalized for the first time by the Act of 1824; but this was on the recommendation of the Commissioners (appointed in 1818), who considered this 10-lb. rating to be in agreement 'with the standard pint in the Exchequer, which is found to contain exactly twenty ounces of water.' As this Exchequer pint was an Elizabethan standard dated 1602, it is quite certain that our present standard gallon was intended to perpetuate ancient tradition.

In 1758 the volume of this pint was reported * to be 34·8 c. in., and that of a quart dated 1601 was found to be 70 c. in.; but in 1688 its volume was recorded as 70·5 c. in., and in 1700 the proportional 4 such quarts (= 282 c. in.) was claimed, by the importer who had declared a consignment on that basis, to be the Exchequer wine gallon. The prosecution contended that the quart was for ale, not wine—an assertion founded on Acts passed in the reign of Charles II (presumably to enforce tradition) requiring ale and beer to be gauged by the quart. A sharp distinction, therefore, must be drawn between a gallon proportional to 4 qts. dated 1601 and the actual Elizabeth gallon measure of that date. There were two such measures at the Exchequer in 1688 and they were found † to be equal to the Henry VII gallon, with a volume of 272 c. in.: evidently they were copies of the older standard, which means that the Elizabeth gallon measure of 1601 must be regarded as a Winchester corn gallon. The quart of that date, however, was intended for ale and (on the evidence of the pint dated 1602) its capacity rating should be 40 ounces: this makes the proportional 4-qt. standard a 10-lb. gallon and a prototype of our present standard gallon.

It is convenient and should not be confusing to call this 4-qt.

---

* By the Parliamentary Committee appointed 'to inquire into the original standards of weights and measures in this Kingdom and to consider the laws relating thereto.'

† By the Commissioners of Excise in their memorandum to the Lords of the Treasury, dated 15 May 1688.

standard the Exchequer gallon; and it should be noted that if it is rated $4 \times 70 = 280$ c. in. it will be exactly in (10/8) volume ratio to the Guildhall gallon of 224 c. in. and will give this smaller gallon an 8-lb. rating.   Moreover, both gallons will then be exactly proportional to the Excise wine gallon rated 10 troy lb. if the volume of that gallon is reckoned as $8 \times (1/60)$ c. ft. $= 230.4$ c. in.   It seems reasonable, therefore, to suppose that the following table reflects medieval intention:

Guildhall (gallon)	Excise (gallon)	Exchequer (4 quarts)	
8 lb.	10 troy lb.	10 lb.	
224	230·4	280	c. in.

The common density at which these relationships of mass and volume are valid is 250 gt./c. in.   The density of water at 62° F. is 252·286 gt./c. in., corresponding to a 10-lb. gallon of 277·418 c. in., and although the 1878 Act does not give a legal volumetric equivalent for the gallon this is the value currently accepted by the Board of Trade.

Further support for the hypothesis that the 4-qt. standard is properly rated as a 10-lb. gall. is to be seen in the evidence relating to the barrel of beer, and that evidence reveals an antiquity (in a Roman aspect of the metrology) that seems sufficient (when taken in conjunction with the text of the *Tractatus*) to explain the existence of the Guildhall gallon.   There can be no doubt that this measure was intended to represent the gallon of the old laws, but it was an interpretation that ignored any reference to wheat, and the pound used as the basis of its 8-lb. wine-weight rating was averdepois, not troy.

XII(10) THE BARREL OF BEER

In their memorandum to the Treasury, written in 1688, the Commissioners of Excise say they are gauging ale and beer by a quart of 70·5 c. in. in pursuance of 12 Charles II, c. 23 (1660),* which contains this article: †

---

* The year of the Restoration.
† Article 20.   Number and text from *Statutes at Large*.

Be it enacted . . . that every six and thirty gallons of beer taken by the gage according to the standard of the ale quart, four whereof make the gallon, remaining in the custody of the Chamberlain of His Majesty's Exchequer, shall be reckoned accounted and returned by the gager for a barrel of beer; and every two and thirty gallons of ale taken by the gage according to the same standard shall be . . . returned for a barrel of ale; and all other liquors aforesaid according to the wine gallon.

For reasons already explained it seems legitimate to give the 4-qt. ale gallon a 10-lb. rating, and further support for this interpretation is apparent when it is seen to make the barrel of beer contain 360 lb. = 500 librae = sack (*Tractatus*). Also it shows the Scots 12-gall. measure to be equal to this barrel when the Scots gallon is rated 12 ale qts. = 480 oz. = 500 unciae.

A clause in the Act of Union required the English standards to be used for gauging in Scotland, and there is a memorandum (dated 16 October 1707) from the Excise Office at Edinburgh to the Commissioners in London reporting these discrepancies:

We do find that the 12-gallon present Scots standard measure makes 35 gallon and an half English when measured by the English quart but when measured by the English gallon the aforesaid 12 gallons make 37. This difference makes the brewers very obstinate in all matters relating to the Excise.

Presumably this obstinacy arose from the Excise trying to collect duty on a 37-gall. rating for the Scots standard measure, and if so it was well justified because the only English gallon measure that the Excise could have used to give this rating was the Winchester corn gallon and even this must have been less than 272 c. in. if the recorded measurements were exact because 37 such gallons would exceed 35·5 gall. in terms of a 70·5 c. in. quart. This latter measurement is equivalent to 36 gall. of 278 c. in., but in any case this analysis of the evidence suggests that the Scots measure and the 36-gall. barrel both reflect a tradition in which they may have been rated 50 Roman congii and had a nominal volume of 1 myriad c. in.

XII(11) 'Seal'd Quarts'

From time immemorial * it has been a cardinal point of principle that weights and measures must not only be true, but must also be stamped: for example:

A.D. 978–1016. Ethelred (the Unready). De Institutis Lundonie. 'And those who have charge of towns shall see to it that every weight is stamped according to the standard of my mint.'

A.D. 1066–87. William the Conqueror. De Mensuris et Ponderibus. 'Et quod habeant per universum regnum mensuras fidelissimas et signatas, et pondera fidelissima et signata, sicut boni predecessores statuerunt.'

A.D. 1216–72. Henry III. Statutum de Pistoribus. 'All the measures of every town both great and small shall be viewed and examined twice a year. The standards of bushels, gallons and ells shall be sealed with the iron seal of our Lord the King.'

A.D. 1594. Date assigned to Shakespeare's *Taming of the Shrew*. The bemused Christopher Sly is hoaxed by the lord's servants with a story that he has been asleep for fifteen years.

Sly. But did I never speak of all that time?

Serv. Oh, yes, my lord, but very idle words. . . . And rail upon the hostess of the house; and say, you would present her at the leet, because she brought stone jugs and no seal'd quarts.

By the Weights and Measures Act of 1878, every measure and weight used for trade must be verified and stamped by an inspector; and it is illegal to use any weight or measure that is not of the denomination of a Board of Trade standard. These points of principle and practice emphasize the distinction between the 1864 and 1897 Acts relating to metric units: the former legalized the use of metric metrology in contracts and provided an authorized table of equivalents; but the latter legalized the use of actual metric weights and measures, and required the Board of Trade to include metric denominations among its standards.

---

* A Greek bowl found by the American School of Classical Studies at Athens has a sealed lead plug riveted through its side: the impression on the seal represents Dionysos. (*Hesperia*, 1949, p. 108.) Also see p. 130.

XII(12) THE YARD

In the Westgate Museum at Winchester there is a brass rod of hexagonal section catalogued as a standard yard of the time of Henry I but with re-standardized ends stamped E (for Edward I) and H (for Henry VII) respectively. Its length is recorded as 0·04 in. short. The seals form the actual extremities of the measure: they have flats that overlap corresponding flats on the rod to which they are riveted. The earlier seal (E) is an example of the 'iron seal of our Lord the King' mentioned in the Statutum de Pistoribus of Henry III; the metal of H matches that of the rod.

A yard measure based on the Elizabethan standard in the Exchequer was made for the Royal Society in 1742 and became the principal unofficial standard of reference used by Bird in 1760 when he made a standard yard for the Parliamentary Committee of 1758, and this yard was adopted by the Commissioners appointed in 1818. They remarked in their report of 1820: 'We have found reason to prefer the Parliamentary standard executed by Bird in 1760.'

Winchester City Museums photo of the sealed extremities of the ancient yard in the Westgate Museum.

XII(13) The Pole, Mile, and League

The mile of 1760 yds. received statutory definition in 35 Elizabeth, c. 6 (1592); this was an Act to prevent the erection of new buildings 'within three myles of the gates . . . a myle to conteyne eight furlongs and every furlong to conteyne fortie luggs or poles and ev'y lugg or pole to conteyne sixteen foot and half.'

In the Statutum de Admensuratione Terrae,* where the principal text is a proportional table giving the length and breadth of rectangular acres from 10 × 16 poles to 80 × 2 poles, the final paragraph defines the pole as 16·5 ft.:

> And be it remembered that the Iron Yard of our Lord the King containeth three feet and no more.   And a foot ought to contain 12 inches, by the right measure of this yard measured; to wit the thirty-sixth part of this yard rightly measured maketh an inch neither more nor less.   And five yards and a half make one perch, that is sixteen feet and a half, measured by the aforesaid Iron Yard of our Lord the King.

That a pole of 16 ft. was substituted for that of 16·5 ft., and used as the basis for a league of 12 furlongs (480 poles), is evident from this reference quoted by Morgan: †

> Leuga autem Anglica duodecim quarentenis conficitur. Quarentena vero quadraginta perticis. Pertica habet longitudinis sedecim pedes.

This league survived as the traditional Cheshire mile of 2560 yds. = 3 × 16 × 480 ft.; and this evidently was the mile in terms of which William of Worcester (c. 1478) gave 12 miles instead of 17·5 miles as the distance from Oxford to Farringdon.‡

In Domesday, according to Morgan, the league of 12 furlongs is in common use and the mile is rarely mentioned.   On the Gough (Bodleian) map of England (c. 1335) the marked distances, according to Petrie,§ are all in units that exceed the

---

* Of uncertain date but sometimes attributed to 33 Edward I (1305).
† *England under the Norman Occupation*, by J. F. Morgan (1858).
‡ *Itinerarium Willelimi de Worcestre*, by Dr. Nasmyth (1778).
§ *Proceedings of the Royal Society of Edinburgh* (1883).

statute mile; his examination suggested the use of three different standards:

> 2220 to 2240 yards for most of England
> 2499 ,, 2657   ,,   ,, Wales
> 2534 ,, 2886   ,,   ,, Cheshire and Shropshire

It is apparent that the lower limit of the range for Cheshire is nearly equal to the Cheshire mile, while the upper limit is about one-eighth longer; within this range it is possible that the league of 12 statutory furlongs (= 1·5 miles = 2640 yds.) was used, and this could be accommodated also within the limits of the range for Wales. In England, however, the league seems to have been 10 furlongs (= 1·25 miles) = 2200 yds., although the upper limit happens to be equal to the Irish mile.

XII(14) GEOMETRIC ACRES

In addition to the statutory English acre of 4840 sq. yds., there was a Scottish acre of 6150·4 sq. yds. and an Irish acre of 7840 sq. yds.; but they all measured 160 sq. poles in terms of the local pole, and 40 of these made the local furlong that was one-eighth of the local mile. These furlongs are 220, 248, and 280 yards in length respectively; which makes the (English/Irish) furlong length ratio = $(22/28) = (\pi/4)$, for $\pi = (22/7)$. The (English/Irish) acre ratio, therefore, is $(\pi/4)^2$; and the (English/Scottish) acre area ratio is as near to $(\pi/4)$ as could be obtained with a furlong measured in integral yards and an acre defined as 1 furlong × $(1/10)$ furlong. Thus, the Scottish acre is the geometric mean in a series having $(\pi/4)$ as its common factor.

These ratios imply that the English and Scottish acres can be represented geometrically by circles inscribed respectively in square Scottish and Irish acres. And if the Irish acre is increased by 14 sq. yds. to 7854 sq. yds. it would measure $(\pi/4)$ myriad sq. yds. and be equal to the area of a circle 100 yds. in diameter: this is the published diameter of the outer earthwork circle at Stonehenge. A square enclosing this circle would have an area of 1 myriad sq. yds. and be equal to the Hindu nivartana. This geometry of the acres, in conjunction with that of the Great

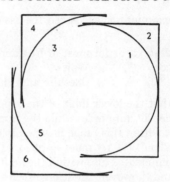

THE GEOMETRIC ACRES

(1) English circular acre in (2) Scottish square acre.
(3) Scottish   ,,   ,, ,, (4) Irish   ,,   ,,
(5) Irish   ,,   ,, ,, (6) Hindu   ,,   nivartana.

Pyramid, suggests that the geometers of remote antiquity may have realized $\pi = (22/7)$ to be the numerical value of (area of circle/radius^2) as well as of (circumference/diameter).

The area of the English acre is $4840 \times 1296 = 6272640$ sq. in. $= (6272640/1550) = 4046 \cdot 86$ sq. metres, which is only 1 part in 600 less than the area of the geodetic acre represented by $(100A)^2$ $= (63 \cdot 662)^2 = 4052 \cdot 85$ sq. metres and defined as one-myriad-millionth of the square on the radius of the spherical Earth. If these acres are represented by squares, their sides are $2504 \cdot 5$ and $(63 \cdot 662 \times 39 \cdot 37) = 2506 \cdot 373$ in. respectively; the linear difference is less than 2 in., that is less than 1 part in 1200. The combined evidence of the English acre and Palestinian field suggest that the geodetic acre was the prototype of both.

As a square, the English acre has a side that is incommensurate with the English linear scale. As a circle its appropriate rating would be that of the geodetic acre expressed as 2 million $\pi = 6283185$ sq. in., which implies a radius of $1000\sqrt{2}$ in.: this, hypothetically, was the length of a linear unit called ying in ancient Chinese metrology.

The area ratio of the English acre to the Roman jugerum is exactly $(8/5)$ if the jugerum is reckoned as equivalent to the square on 96 royal cubits rated 20 + (5/8) in. If the English acre is identified with the geodetic acre, then the radius of the spherical earth can be rated 100000 acre-sides in terms of the side of this acre regarded as a square. This acre-side measures $96\sqrt{(8/5)}$ royal cubits, and if the royal cubit is now rated $20\sqrt{2}$ digits in terms of a digit defined as (1/54) metre, then the radius and the circumference of the spherical earth are linked by this geometric metrology. In terms of the digit defined as (1/54) metre, the circumference measures 2160 million digits: in terms of the acre-side, it measures $1536\pi/(\sqrt{5})$ million digits. If 1·4062 is written for $\pi/(\sqrt{5})$ these expressions are equal, but a more accurate value for this ratio would be 1·405.

In Scotland the local pole, called fall, was divided into 6 ells; the Scottish acre, therefore, measured 5760 sq. ells in terms of an ell of 37·2 in. = 3·1 ft. The English ell measured (5/4) yard = 45 in. The French ell (aune) measured 46·656 in., that is 4 Roman ft.; hypothetically, this was the intended length of the Aeginetan iron currency spit.

XII(15) WEIGHTS AND MEASURES IN CURRENT USE

Inches	Feet	Yards	Furlongs	Mile
63360 =	5280 =	1760	= 8 =	1
7920 =	660 =	220	= 1	
36 =	3 =	1		
12 =	1			

Inches	Poles	Furlong	Chains	Links
	40 =	1 =	10 =	1000
198	= 1			
7·92 =				1

Sea mile = 6080 ft.   (Admiralty standard.)
　　　　　6080 ft./hour = 1 knot.   (Unit of speed at sea.)

Fathom = 6 ft.   (Unit of depth at sea.)
Hand　 = 4 in.   (Unit of height, for horses.)
Nail　 = 2·25 in. = (1/16) yd.   (Used in cloth measure.)

$$\begin{array}{ccc} Sq.\ in. & Sq.\ ft. & Sq.\ yd. \\ 1296 & = \quad 9 \quad = \quad 1 & (1296 = 6^4 = 1000k) \\ 144 & = \quad 1 \end{array}$$

$$\begin{array}{cccccc} Sq.\ yds. & Sq.\ poles & Roods & Acre & Sq.\ chains & Sq.\ links \\ 4840 & = \quad 160 \quad = & 4 \ = & 1 \ = & 10 & = 100000 \\ & & & & 1 & = \ 10000 \end{array}$$

The surveyor's chain of 100 links (= 4 poles = 66 ft.) was invented by Edmund Gunter (1581–1626) and designed to decimalize the acre.

$$\begin{array}{ccc} C.\ in. & C.\ ft. & C.\ yd. \\ 46656 & = \quad 27 \quad = & 1 \\ 1728 & = \quad 1 \end{array}$$

Lb.	St.	Qr.	Cwt.	Ton	
2240 =	160 =	80 =	20	1	St. = stone = 2 cloves
112 =	8 =	4 =	1		Qr. = quarter
14 =	1				Cwt. = hundredweight
100 =	'Cental' (legalized in 1879.)				

Lb. = pound (averdepois)

Grains	Oz.	Lb.	
7000	= 16 =	1	This is the only legal pound.
437·5	= 1		Dram = (1/16) oz. Obsolete?

The troy ounce of 480 gt. is legal for certain purposes and is the basis on which the price of gold is quoted.   In the old money scale it was divided into 20 pennyweights but in the apothecaries' system its division is into 8 drachms or 24 scruples.   The apothecaries' fluid-ounce of 8 fluid-drachms or 480 minims, however, is the averdepois ounce of water; that is (1/20) pint.

Pint	Gallon	Bushel	Quarter	
512 =	64 =	8 =	1	Chaldron = 36 bushels
64 =	8 =	1		Bushel = 4 pecks
8 =	1			Gallon = 4 quarts
				Pint = 4 gills

Sale by weight has rendered virtually obsolete the dry measures chaldron, quarter, bushel, and peck.

Gallon = Capacity for 10 lb. of water at 62° F. when weighed under specified conditions. There is no statutory volumetric equivalent, but the Standards Department of the Board of Trade currently accepts 277·418 c. in.   This corresponds to a density of 252·286 gt./c. in., which is considered to express the best available evidence relating to the density of water under the specified conditions.

Reputed quart = (1/6) gall. Customary capacity of a wine bottle.

Litre = Capacity for 1 kgm. of water at 4° C. Volumetric equivalent currently accepted is 1000·028 c.c.

Metric carat = 200 mgm. = 3·08646 gt.   This is the only carat weight now in customary use. The London carat weight was (1/151·5) troy oz. = 3·16831 gt.

Metric equivalents.   (S.R.O. 411 of 1898.)

    1 yd. = 0·914399 metre   1 metre = 1·0936143 yds.
    1 lb.   = 0·45359243 kgm.   1 kgm. = 2·2046223 lb.
    1 gallon = 4·5459631 litres

Approximate values

    1 in. = 25·4 mm. (legal for use in trade).
    1 sq. in. = 6·45 cm.^{2} 1 c. in. = 16·4 cm.^{3}
    1 acre = 4047 m.^{2}
    1 gt. = 0·648 gm.   1 lb. = 453·6 gm.   1 gall. = 4·5 litres.
    1 metre = 39·37 in. (legal in the U.S.).   1 km. = (5/8) mile.
    1 m.^{2} = 1550 sq. in.   1 m.^{3} = 35·3 c. ft.
    1 litre = 0·22 gall. = 61 c. in.
    1 gm. = 15·4 gt. 1 kgm. = 2·2 lb. 1 gm./cm.^{3} = 253 gt./c. in.

Some links with antiquity (also see XII(5)):

English	Sumerian	English	Roman
Pole =	10 cubits	Acre	= (8/5) Jugerum
Sq. pole =	100 sq. cubits	Tun (wine)	= 36 amphorae
Acre =	16000 sq. cubits	Barrel (beer)	= 50 congii
Furlong =	600 ft.	70 troy lb.	= 80 librae
Mile =	3200 cubits	24 oz.	= 25 unciae

    Sumerian c. ft. = 10 troy gall.
    Entemena's vase = 10 ,, pints
    Assyrian c. ft. = Winchester bushel
    Euboic mina = Tower lb. = (15/16) lb. troy

XIII

Weights and Measures of the United States of America *

U.S. gallon = 231 c. in.   This is the standard measure for liquids and is the only gallon in U.S. metrology.   Its volume is that of a cylinder 7 in. diameter by 6 in. deep, and its prototype was the wine gallon legalized in 6 Anne, c. 27 (1706).   When, by the Act of 1824, this gallon became obsolete in England, H.M. Customs and Excise was authorized to use (5/6) as its legal ratio to the new 10-lb. gallon.   This implies a capacity for 8 + (1/3) lb. of water, but the capacity corresponding to the U.K. gallon rated 277·418 c. in. is 8·32318 lb.

U.S. bushel = 2150·42 c. in.   This is the standard dry measure and is the only bushel.   Its volume is that of a cylinder 18·5 in. diameter by 8 in. deep, and its prototype was the cylindrical version of the Winchester bushel legalized in 8 William III, c. 22 (1696).

U.S. quarts and pints.   It is necessary to define these as liquid or dry.

U.S. liquid quart = (1/4) U.S. gallon = (5/6) U.K. quart †
  ,,   ,,   pint = (1/8)   ,,   ,,   = 28·875 c. in.
                    water-weight = (25/24) lb. = 16 + (2/3) oz.†
  ,, dry   ,,   = (1/64) U.S. bushel = 33·6003 c. in.

U.S. fluid oz. = (1/16) U.S. liquid pint = 1·8046875 c. in.
              = 8 U.S. fluid drams = 480 U.S. minims.
              water-weight = (25/24) oz. †
              ∴ 24 U.S. fl. oz. = 25 U.K. fl. oz. †

---

* This summary is confined to principal differences between the U.S. and U.K. metrologies.   For further information see Letter Circulars LC 449, 681, and 682 of the National Bureau of Standards (U.S. Department of Commerce), Washington.
† Approx.

U.S. yard and pound.   These are the same as the corresponding U.K. units, but the U.S. standards are defined in terms of the metre and the kilogram thus:

$$\text{(U.S. yard/metre)} \ \ = (3600/3937)$$
$$\text{(U.S. pound/kilogram)} = (10000/22046)$$

This means that the U.S. legal standard inch is of such a length that there are exactly 39·37 of them in the metre.   Expressed in decimals, the U.S. standards are:

U.S. yard = 0·9144018 metre.   U.S. inch = 25·4000508 mm.
U.S. pound = 453·5924277 gm.

The above definitions date from the Mendenhall order of 1893: the use of the metric system was legalized in 1866.

U.S. hundredweight and ton.   Used without qualification, hundredweight and ton commonly mean 100 lb. and 2000 lb. respectively: when misunderstanding is possible these units are called short or net in order to distinguish them from the long or gross hundredweight of 112 lb. and ton of 2240 lb.   Prices sometimes are quoted per gross ton and per 100 lb. for the same commodity. It is customary to specify weights in pounds even when far in excess of a ton.

U.S. troy pound = (5760/7000) U.S. lb.   This definition dates from the Mendenhall order of 1893 and refers the troy pound indirectly to the kilogram.   Previously the averdepois pound was defined in terms of the troy pound of the Mint, legalized in 1828, but since 1911 the standard for coinage purposes has been the troy pound of the National Bureau of Standards.

U.S. assay ton (AT) = 29·167 gm.   Sample mass in an assay.

Milligrams/(AT) = troy oz./ton of 2000 lb.

U.S. cup = (1/2) U.S. liquid pint.   A customary cookery measure.

U.S. sea mile = 6080·2 ft.

U.S. time zones.   Eastern, Central, Mountain, and Pacific. The time in these zones is slower than Greenwich time by 5, 6, 7, and 8 hours respectively.   The difference in longitude corresponding to 1 hour is 15 degrees.

XIV

Appendices

XIV (1) POWERS OF 6 AND k

These are the first twenty powers of 6:

Power	Number	Power	Number
1	6	11	362 797 056
2	36	12	2 176 782 336
3	216	13	13 060 694 016
4	1 296	14	78 364 164 096
5	7 776	15	470 184 984 576
6	46 656	16	2 821 109 907 456
7	279 936	17	16 926 659 444 736
8	1 679 616	18	101 559 956 668 416
9	10 077 696	19	609 359 740 010 496
10	60 466 176	20	3 657 158 440 062 976

These approximations to powers of 6 have a constant error of $+ (1/4375)$ exactly:

Power	Approximation	
4	$(35/27) \times 10^3$	$= 1 \cdot \dot{2}9\dot{6} \times 10^3$
5	$(7/9) \times 10^4$	
7	28×10^4	$\therefore (1/4)6^7 = 7 \times 10^4$
8	168×10^4	$= (42/25) \times 10^6$
9	1008×10^4	

These are the first five powers of k:

k	k^2	k^3	k^4	k^5
$(6^4/10^3)$	$(6^8/10^6)$	$(6^{12}/10^9)$	$(6^{16}/10^{12})$	$(6^{20}/10^{15})$
1·296	1·679 ..	2·176 ..	2·821 ..	3·657 ..
1·296	1·68	(Approx.)		3·657 ..
$(35/27)$	$(42/25)$			$(128/35)$
$+ (1/4375)$		(Error)		$- (1/230000)$

XIV(2) Measurements of the Old Athena Temple

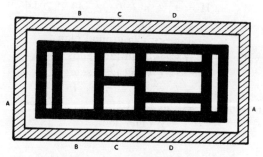

P = Published measurements in metres.*
H = Hypothetical intended measurements.
( ) = Wall thicknesses.   Gk. = Greek.

Floor divisions (west to east) along line AA:

P = 1·55(1·55)6·20(1·20)6·20(1·35)10·5(1·60)1·40 = 31·55
H = 1·50(1·50)6·25(1·25)6·25(1·25)10·5(1·50)1·50 = 31·50

Floor divisions (south to north) along lines:

	BB	CC	DD
P =	10·65	4·85(1·25)4·50	1·8(1·30)4·35(1·30)1·9
H =	10·66	4·69(1·28)4·69	1·9(1·27)4·32(1·27)1·9

Cella

(external) P = 34·7 × 13·45
H = 34·72 × 13·44 = 112·5 × 43·5 Gk. ft.
= 4900 Gk. sq. ft. = 100 × 7^2 Gk. sq. ft.

(floor) P = 31·55 × 10·65
H = 31·5 × 10·66 = 102·08 × 34·56 Gk. ft.
= 3528 Gk. sq. ft. = 72 × 7^2 Gk. sq. ft.

---

* *The Acropolis of Athens,* by Martin L. D'Ooge (1908).

Stylobate

(external) $P = 43\cdot44 \times 21\cdot34$ |
$\qquad\qquad 43\cdot95 \times 21\cdot85$ |

$\qquad H = 43\cdot2 \times 21\cdot6 = 140 \times 70$ Gk. ft.
$\qquad\quad = 2 \times 4900$ Gk. sq. ft.
$\qquad\quad = 2 \times$ Cella ground plan

(wall) $\quad P = (2\cdot25)$ and $(2\cdot1)$ ends
$\qquad\qquad (2\cdot1\ )$ and $(2\cdot1)$ sides

$\qquad H = (2\cdot16)$ m. $= 7$ Gk. ft.

(internal) $H = 126 \times 56$ Gk. ft. $= 2 \times 3528$ Gk. sq. ft.
$\qquad\quad = 2 \times$ Cella floor.

XIV(3) MEASUREMENTS OF THE CONICAL CUPS

Published measurements	*Silver*	*Bronze*	
D = Inside diameter (top)	109	72·8	mm.
d = ,, ,, (bottom)	30	26·2	mm.
L = Length of slant side	169·3	100·1	mm.

The following calculation, based on the above measurements, shows that both cups were designed with a common cone angle $\theta = 27°$. Thus:

$$\mathrm{Sin}(\theta/2) = (D–d)/2L = 0\cdot233 \text{ for silver and bronze}$$
$$= \sin 13° 30' \quad \therefore \ \theta = 27°$$
$$\mathrm{Cot}(\theta/2) = (100/24)$$

Length of slant side of cone, between base of cup and apex:
$$a = (d/2)/\sin(\theta/2) = 64\cdot4 \text{ mm. (silver)}; 56\cdot3 \text{ mm. (bronze)}$$

Volume of the geometrical apex cone, below the base:
$$q = (d/2)^3 (\pi/3) \cot(\theta/2) = 14\cdot7 \text{ c.c. (silver)}$$
$$= 9\cdot8 \text{ c.c. (bronze)}$$

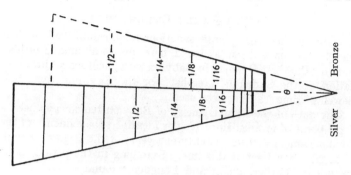

Published positions of the rings
(millimetres from the base along the slant side)

S = Silver; B = Bronze.

                (1/2)   *dja   hin   top*
S = 3·8 8·3 13·8 21·8 35·8 55·3 82·3   105·3 119·8 147·3 169·3
B = 9·0 14·1 20·8 34·8 52·6 72·5 100·1 (broken)

At level of any calibration ring, let:

b = slant length as shown above, and $C = (a + b)$
Q = volume of cone (including q) $= (C/a)^3 q$
$V =$,,   ,, cup $= (Q - q)$
$J =$,,   ,, dja as indicated by V multiplied by the denominator of the fraction to which it relates:

	(1/16)	(1/8)	(1/4)	(1/2)	1	dja
Silver J =	331	326	322	320	331	c.c.
Bronze J =	509	490	462	400	—	c.c.

The volumes of the dja, deduced from the volumes at (1/8) dja level, are:

Hypothetical: c.c.   Water-weight
   Dja (silver) = 326·67   Roman libra = 12 unciae
   ,, (bronze) = 490   French livre = 18 ,,
   Hin (silver) = 510   18 oz.

XIV(4) Early Chronology

Dates earlier than those sanctioned by established ancient chronology have been assigned here by combining Smith's [1] date for the First Dynasty of Babylon with Jacobsen's [2] interpretation of the Sumerian king-list: for the Egyptian dynasties, Sewell's [3] time scale has been used.

The date of the First Dynasty of Babylon has always been a focal point of interest since Kugler identified the evidence of the Venus Tablets with the twenty-one years of the reign of Ammizaduga, tenth king of this line. Searching in the period 2000–1800 B.C., Fotheringham and Langdon [4] found a sequence of twenty-one years in which the astronomical facts relating to the rising of Venus agree with the records of the Tablets; and this discovery led Langdon to propose 2169 B.C.[5] as the starting date for this dynasty; but Sewell (at the instigation of Smith) subsequently calculated that an equally appropriate series occurred 275 years later, and Smith has shown that the evidence revealed by excavations at Alalakh requires this revision: his date for the beginning of the First Dynasty of Babylon is 1894 B.C.

Of several extant Sumerian king-lists, the most complete is that presented by Weld-Blundell to the Ashmolean Museum and sometimes called the Oxford Prism; its text begins with an antediluvian reference:

> When kingship was lowered from heaven, the kingship was in Eridu: in Eridu, Aulim became king and reigned 28800 years. (Jacobsen.)

After eight kings had reigned for 241200 years in five different capitals:

> The flood [6] swept thereover. After the flood had swept thereover, when the kingship was lowered from heaven, the kingship was in Kish.

Kish was the principal city in Akkad; but when it was 'smitten with weapons' the dynastic capital moved south to Erech in Sumeria, then further south to Ur and later back to Kish and so on; and if all these early dynastic periods are recorded in sequence it is necessary to assign some such date as 5500 B.C. to the beginning of the first in order to accommodate the rest in advance of

THE OXFORD PRISM

List of the kings who ruled Mesopotamia from before the Flood to the Dynasty of Isin, *c.* 2000 B.C. Width across each of the four faces about 4 in.; height about 8 in. Ashmolean Museum photo of exhibit 1923.444.

established history.   When an inscription revealed Mesannipadda (founder of the First Dynasty of Ur) under conditions showing that he could not have reigned so long ago as was thus indicated, students realized the need for some other interpretation.   A reconstruction on the principle of overlapping dynasties was published in 1939 by Jacobsen, and its graduated effect on the earlier hypothetical dates given in the *Cambridge Ancient History* is indicated by the following examples:

	Kish I	Ur I	Kish III	Agade	Ur III	
C.A.H.	5500	4216	3089	2872	2474	B.C.
Jacobsen	3100	2850	2690	2615	2400	B.C.

As Jacobsen uses the Langdon-Fotheringham date for the First Dynasty of Babylon, I have subtracted 275 years from his date for Kish III and subsequent dynasties: it should be noted, however, that whereas his date makes Kish III contemporary with Sewell's date for the Pyramid Age (Dynasty IV) in Egypt, my arbitrary change makes it contemporary with the end of the Old Kingdom in Dynasty VI.   I do not need to go farther back than Kish III in order to cover the Babylonian metrological monuments; but in Egypt the Pyramids themselves provide evidence.

Egyptian chronology is based on the hypothesis that events were recorded in terms of a continuous calendar of 365-day years that was instituted on a day when the heliacal rising of Sirius [7] coincided with the summer solstice [9]: on the evidence of a passage in *De Die Natali*,[8] A.D. 139 is taken as a key date for the backward reckoning of such Sothic cycle starting-points:

A.D. 139; $(1460-139) = 1322$ B.C.; $(1322 + 1460) = 2782$ B.C.[9]

As the heliacal year is 365·25 days in length; [10] a Sothic cycle of $4 \times 365 = 1460$ such years covers 1461 years of 365 days each, and a heliacal rising marking the beginning of an intermediate year in the cycle would occur on a date in the current 365-day year that would indicate an elapsed fraction of that year proportional to the elapsed fraction of the cycle.

One such event is recorded as having occurred during the seventh regnal year of Senusret III on a day that is interpreted as the 226th day [11] of the current 365-day year; which implies that $4 \times 226 = 904$ years of the Sothic cycle that began in 2782 B.C.[12]

had elapsed, and that Senusret's seventh regnal year occurred in (2783−904) = 1879 B.C. As this regnal year is known to have been the 120th dynastic year, the Twelfth Dynasty came to power with the accession of Amenemet I in 1998 B.C. Sewell gives 1990 to 1777 for this dynasty, and 1573 to 1314 for the Eighteenth Dynasty: during much of the intervening two centuries, the Hyksos were in the land.

Two other heliacal risings are on record; one (mentioned in the Ebers Medical Papyrus) is associated with the ninth year of Amenhotep of the Eighteenth Dynasty, the other associates the cycle beginning in 1322 with Ramses of the Nineteenth Dynasty: this is on the evidence of a reference (by Theon the mathematician) to this cycle as the epoch of Menopheres, which is assumed to be a reference to Ramses's throne name Menphere.

It is from the combination of key dates such as these with monumental and other evidence relating to regnal years and events, that tentative Egyptian chronologies have been compiled; but no such time scale can stand alone because the evidence also includes certain points of synchronism such as those revealed in the Tell-el-Amarna letters. The earlier the date, the greater its uncertainty; and the more likely are radical changes of opinion to occur when new evidence is revealed by archaeological research.

The Egyptian Dynasties

			The Old Kingdom			*Interregnum*
I	II	III	IV	V	VI	VII–X
3189	3142	2816	2691	2556	2421 B.C.	

			The Middle Kingdom				*The Empire*	
XI	XII	XIII	XIV	XV	XVI	XVII	XVIII	XIX
	1991	1778			Hyksos		1574	1315

The Decline
XX and after
1135 ,, ,, B.C.

The relationship in time here assumed between the Babylonian and Egyptian dynasties from 2500 to 1500 B.C. is shown graphically on page 186.

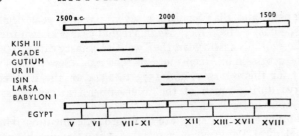

REFERENCES AND NOTES

1. *Alalakh and Chronology*, by Sidney Smith (1940).

2. *The Sumerian King List*, by Thorkild Jacobsen (1939).

3. 'The Calendars and Chronology,' by J. W. S. Sewell. A chapter in *The Legacy of Egypt*. Edited by S. R. K. Glanville (1942).

4. *The Venus Tablets of Ammizaduga*, by S. Langdon and J. K. Fotheringham (1928).

5. The date previously published in the *Cambridge Ancient History* was 2225 B.C.

6. When Woolley found 8 ft. of clay silt submerging the earliest rubbish of human occupation at Ur, it was evident that there had been a really devastating flood in Sumeria.

7. It is believed that the Egyptians watched regularly for the heliacal rising of the brightest of all stars, Sirius; the greater dog-star that they called Sothis. When a star rises with the Sun it is invisible and remains so; all day it is lost in the Sun's rays and all night it is below the horizon; but the interval (sidereal day) between its successive crossings of the meridian being nearly 4 minutes less that the mean solar day of 24 hours by the clock, there comes the day of its heliacal rising—the first day in each year on which it is momentarily visible in the darkness preceding the dawn.

8. 'Sed horum initia semper a primo die mensis ejus sumuntur cui apud Aegyptios nomen est Thouth, quique hoc anno fuit ante diem VII Kal. Iul., cum abhinc annos centum imperatore Antonino Pio II Bruttio Praesente Romae consulibus idem dies fuerit ante diem XIII Kal. Aug., quo tempore solet canicula in Aegypto facere exortum.' (Censorinus, *De Die Natali*, 21. 10.)

The reference to 20 July (Gregorian), as the day on which the dog-star in Egypt was wont to make its rising, is read to mean that a Sothic cycle began in A.D. 139—the year of accession of Emperor Antoninus Pius, when (for the second time) he became consul; and Bruttius Praesens was elected to the other consulship.

9. Sewell states (p. 4) that the day of the heliacal rising and the day of the summer solstice were virtually coincident (about 17 July, Julian) at this (2782 B.C.) epoch; and he suggests that Imhotep, minister and architect to Zoser of the Third Dynasty, introduced the calendar at the beginning of this Sothic cycle.

10. Schoch (*Die Länge der Sothisperioden beträgt 1456 Jahre*) gives 365·2507 days as the mean interval between heliacal risings of Sirius, as seen from the latitude of Memphis. Quoted by J. K. Fotheringham in an article on the calendar published in the *Nautical Almanac* (1931). The length of the sidereal year is 365·2564 mean solar days. The tropical or equinoctial year is 365·2422 mean solar days; and 1505 such years = 1506 years of 365 days.

11. In the *Cambridge Ancient History*, vol. i, p. 168, this date is given as the first day of the month Pharmouthi. Sewell gives it as the 16th day of the 8th month; which is the (7 × 30) + 16 = 226th day of a year of 12 consecutive 30-day months followed by 5 epagomenal days.

12. This is the only appropriate cycle for the Twelfth Dynasty if the calendar was instituted during the Third Dynasty.

XIV(5) The Calendar

The average length of the Julian year being 365·25 days, while the mean duration of the solar year is only 365·2422 days, there was a residual error in the Julian calendar; causing the vernal equinox to become progressively displaced from 25 March, to which date it had been restored by Julius Caesar when he reformed the Roman calendar in 46 B.C.   In order to eliminate accumulated errors, Pope Gregory XIII announced that 5 October should be called 15 October in the year 1582; and this was done in Italy, France, Spain, and Portugal.   Several other Roman Catholic countries adopted the new 1 January 1583 as the beginning of a new year; but the Protestant states of Germany, the Netherlands, and Denmark did not change until 1700.   In England the change took place on 3 September 1752; this day being called 14 September, in accordance with the provisions of the Calendar Act passed in the previous year.   Its preamble reads:

> 1751.   24 George II, c. 23.
> An Act for regulating the commencement of the year; and for correcting the Calendar now in use.
> Whereas the legal supputation of the year of Our Lord in that part of Britain called England, according to which the year begin-neth on the twenty-fifth day of March, hath been found by experience to be attended with divers inconveniences . . . as it differs . . . from the common usage . . . and whereas the Calendar . . . commonly called the Julian Calendar, hath been discovered to be erroneous, by means whereof the Vernal or Spring Equinox, which at the time of the General Council of Nice in the year of Our Lord three hundred and twenty-five, happened on or about the twenty-first day of March, now happens on the ninth or tenth day of the same month . . . may it, therefore, please your Majesty . . .

It was during the twelfth century in England that the Church adopted the Annunciation of the Blessed Virgin Mary (Lady Day, 25 March) as the beginning of the ecclesiastical year: by the fourteenth century it had displaced Christmas Day, in common regard, as the beginning of the civil year; and this status it retained (with diminishing popularity at last) until 1 January displaced it under the authority of the above Act.

This Act, however, specifically maintained the time periods of current financial contracts, which became enforceable, therefore, eleven days after the calendar dates on which they were nominally due.   In so far as such contracts matured on quarter days, the postponed dates for enforcement were:

Quarter Day	Enforceable on
Michaelmas (29 Sept.)	10 Oct.
Christmas (25 Dec.)	5 Jan.
Lady Day (25 March)	5 April
Midsummer (24 June)	5 July

Specified as the terminal date of a taxation period in 39 Geo. III, c. 13, 1798, 5 April has remained the last day of the Income Tax Year; but the Government's Financial Year begins on 1 April. Prior to the Calendar Act, it ended at Michaelmas; but this terminus became 10 October in 1752 and so remained until 1800 when it was changed to 5 January, and this was observed until 1855.

The Gregorian Correction

In order to reduce future residual error, the Gregorian reform includes a provision that century years will not be leap years unless their numbers are divisible by 400: this eliminates three days from every period of 400 Julian years, and reduces the residual error to three days excess in 10000 years.

XIV(6) THE STERLING

The silver penny that acquired the name sterling was introduced by Offa of Mercia, who defeated the men of Kent at Otford in 774. He copied the general form of the new denarius (denier) that Pippin the Short had introduced in France: it was thinner and larger in diameter than the sceatta that hitherto had been the silver coin in general circulation, but on the evidence of specimens in the British Museum it seems likely that Offa maintained the same standard of mass and that this was (1/20) ore = 20 gt.   Under

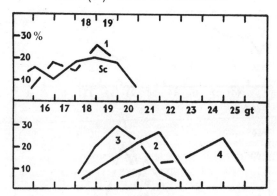

Sc. Sceattas. (1) Offa of Mercia, A.D. 757–96. (2) Mercia (excluding Offa), 768–874; Kent, 765–825; East Anglia, c. 760–890; Northumbria, 877–94. (3) Wessex, 802–71. (4) Alfred (Wessex), 871–901, to Eadwig, 941–59.

	14	15	16	17	18	19	20	21	22	23	24	25 gt.*
Sc.	17	27	19	30	33	32	11					
(1)	3	3	9	7	13	10	2	3				
(2)	2	8	21	52	122	198	298	334	143	52	10	1
(3)	1	3	1	12	39	57	43	18	8	5	1	
(4)	3	7	11	15	29	46	95	122	130	175	228	75

his successors, however, the weight increased until it reached a peak of (1/15) ore = 24 gt. in Alfred's reign. In one of the laws of Ethelred II the pound of the mint is defined as 15 ores, and the peak frequency of his coins suggests that they may have been shorn at 250 to the pound tower.

With the advent of Cnut the penny reverted to the mass category of the sceatta, but it regained weight under Edward the Confessor, and his standard (apparently 20 gt.) was maintained in the first five issues of William I. The evidence relating to the Conqueror's later issues includes that of a hoard of 'Paxs' type pennies found in almost mint condition at Beaworth in Hampshire.

---

* This frequency table is based on the weights of individual coins published in the British Museum Catalogue. All weights from 21 gt. (inclusive) to 22 gt. (exclusive) are tabulated as 21 gt. but plotted as 21·5 gt.

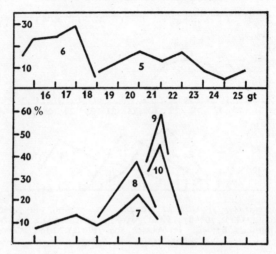

(5) England, Edgar, A.D. 959–75, to Ethelred II, 975–1016.   (6)
Cnut, Harold I, Harthacnut, 1016–42.   (7) Edward the Confessor,
1042–66, and Harold II, Jan.–Oct. 1066.   (8) William I, 1066–87:
series 1–5.   (9) Series 6–8.   (10) William II, Henry I, Stephen.

	14	15	16	17	18	19	20	21	22	23	24	25 gt.
(5)	4	5	13	23	40	75	93	79	96	51	25	49
(6)	58	155	162	201	31	9	20	16	11	4	1	
(7)	76	100	161	192	131	203	344	165	36	6	14	30
(8)	3	9	13	22	41	91	142	48	2			
(9)		1	3	5	10	27	187	459	34	2		
(10)	5	3	9	19	41	60	175	352	90	8		

As 60 per cent of them peak between 21 and 22 gt. it seems possible
that they were shorn at 21 shillingsworth to the pound (tower) of
bullion in terms of the new 12-penny shilling that became customary in England after the Conquest: if so, then the intended mass
was a little less than 21·5 gt.   Although this standard seems to
have been maintained during the next three reigns, the *Chronicle*
tells us that the fineness was much debased while Henry I was
abroad.

A.D. 1124. All this year was the King Henry in Normandy. ... This same year were the seasons very unfavourable ... so that between Christmas and Candlemas men sold the acre-seed of wheat, that is two seed-lips, for six shillings. That was because corn was scarce; and the penny was so adulterated that a man who had a pound at a market could not exchange twelve pence thereof for anything.

A.D. 1125. In this year sent the King Henry, before Christmas from Normandy to England and bade that all the mint-men that were in England should be mutilated in their limbs. ... And that was all in perfect justice, because that they had undone all the land with the great quantity of base coins. ...

The Trial of the Pyx was instituted to detect such frauds but the first known writ for it dates from 1281.* It provided two tests, one for weight and the other for fineness. The type, or mint mark, enabled coins of the same issue (indenture) to be identified and the moneyer's name on the coin enabled the responsibility for fraud (if any) to be located.

In those cases where the king's indentures to his moneyer are extant, the intended mass of the penny is known exactly and an example of such an indenture is given in XIV(7). In XIV(8) the history of the penny from the reign of Henry III is summarized in its relation to the (gold/silver) value ratio.

XIV(7) Indenture for Coinage

New issues of coinage were ordered by an indenture from the king; and I am indebted to the present authorities at the Royal Mint for the following abstract from one dated 23 February 1472 between Edward IV and William Lord Hastings, the Master: perhaps it was the first issue after the King's restoration.

And the said Master hath undertaken to make Five manner of monies of silver. That is to say, One Piece running for 4 pence sterling which shall be called a Groat, and there shall be 112½ of

---

* The trial still takes place (in the presence of a jury composed entirely of members of the Goldsmiths' Company), but the assay of silver is now confined to Maundy money. See Coinage Act 1870 and Trial of the Pyx Order (in Council) 1947.

such Pieces in the Pound weight, another Piece running for 2 pence which shall be called the Half Groat and there shall be 225 such Pieces in the Poundweight. And the third Piece running for a Penny which shall be called the Sterling, and there shall be 450 such Pieces in the Poundweight above said: and the 4th mony which shall be a Halfpenny, which shall be worth half a sterling, and there shall be 900 such pieces in the Poundweight, and the 5th mony shall be called a Farthing, and shall be worth half a Halfpenny, and there shall be 1800 such in the Pound-weight aforesaid.

And all the said monies of silver so made shall be of the assay of a standard of the Old Sterling, that is to say, in every pound-weight of silver of these monies there shall be 11 oz. 2 dwt. of Fine Silver and 18 dwt. of Alloy.

And every poundweight of the monies of silver aforesaid shall hold in number and be in value 37s. 6d. sterling of all the Pieces aforesaid; of which monies of silver the Warden of these Mints shall take up for our said Lord the King, out of every pound-weight so made, 18d. by number; of every which 18d., the said Master . . . shall have by the hands of the said Warden for the working, coining and other expenses and costs of the said matter about the same 12d. by number, and also of which 18d. . . . . the Comptroller, Clipper of the Irons, Clerk and Usher of the Coins shall have their wages. . . .

And there shall remain toward the Merchant of every pound-weight of silver so coined 36s. sterling.

In this indenture the five pieces of money form a dimidiated series from the fourpenny groat to the farthing. In Elizabeth's reign there was a series beginning with sixpence and ending with the three-farthings piece to which Philip Faulconbridge alludes in his devastating parody of brother Robert's skinny physique:

Madam, if my legs were two such riding-rods, my arms such eel-skins stuff'd; my face so thin that in mine ear I durst not stick a rose, lest men should say, 'Look, where three-farthings goes!' (*K. John* I. i).

This series was distinguished from that beginning with the groat by the presence of a rose behind the queen's head, but this reference in *King John* anticipates events by more than three centuries: when the play was written (1596) the allusion was post-topical by thirty-six years, but the coin was struck (1560) four years before Shakespeare was born and it was never reissued.

The tower pound of bullion under this indenture was made to yield 37s. 6d. in coinage but the merchant received only 36s. The difference of 1s. 6d. was the king's seignorage to cover the expenses of the mint as indicated in the text, and it will be noticed that the indenture specifically maintains the old standard of fineness on which the merchant's trust in the coinage was based. It was about 1453 that Henry VIII decided to finance himself by debasing the silver, and the first step seems to have been the addition of 2 oz. of copper to the pound of sterling. This enabled him to coin 56s. instead of the 48s. that the pound troy * was yielding at that time, and of this he took 8s. as seignorage instead of the 1s. that sufficed before. The merchant received 48s. instead of 47s., but the sterling content of this debased metal was less than 42s. Two years later a pound of copper was added to every pound of sterling and this was sufficient to make 96s., of which the king took 40s., leaving 56s. (worth only 28s. in sterling) for the merchant. Under nine-year-old Edward VI another pound of copper was added and the merchant got 60s. (worth 20s.) out of 144s. Finally in 1552, if Ruding's † table of seignorage is correct, the merchant had 120s. (worth 20s.) out of 288s. obtained from a pound of sterling by the addition of 5 lb. of copper.

XIV(8) THE (GOLD/SILVER) VALUE RATIO
IN ENGLAND

The Ancient Britons had an indigenous coinage but the number of extant specimens is insufficient to justify more than a general statement that there were gold and silver pieces of about 20 and 80 gt. and a gold coin of about 120 gt. There was also a copper coinage now represented by specimens spread over the range from 10 to 67 gt. During the Roman occupation the currency came almost entirely from the Roman mints in Gaul, and when

---

* See XII(4).
† Ruding's *Annals of the Coinage of Great Britain*, 1840.

these closed (with the withdrawal of the garrisons) early in the fifth century Britain was left without money, except that found by chance in hidden hoards.

> A.D. 418. This year the Romans collected all the hoards of gold that were in Britain and some they hid in the earth, so that no man afterwards might find them, and some they carried away with them into Gaul. (*Anglo-Saxon Chronicle.*)

Merovingian coinage began to enter the country after A.D. 500 and a few Anglo-Saxon mints may have been in operation by the time St. Augustine arrived in 597. When a certain traveller hid or lost 100 gold trientes in the neighbourhood of Crondall about A.D. 660, 75 per cent of his coins had been made in England. It was in the autumn of 1828 that two brothers Lefroy, while walking on the heath (near Aldershot) 'saw a little heap of apparently brass waistcoat buttons, lying mixed, but with bright edges just washed bare by the late rains.' * The hoard was in a shallow hole made by a turf-cutter's spade and had escaped his notice. There were 100 pieces in the find (published weight 1988·7 gt.) and 97 of them are now in the Ashmolean Museum. They include a half-finished piece and two pellets ready for coining: all are in mint condition and the following frequency table † suggests $k = 1·296$ gm. $= 20$ gt. as the intended mass; this implies that 270 were shorn from the pound tower or 252 from the libra.

(1) Anglo-Saxon coins in the Crondall hoard. (2) In other collections. (3) Merovingian coins in the hoard.

	12·5	1·26	1·27	1·28	1·29	1·30	1·31	1·32	1·33 gm.
(1)	4	4	3	10	11	19	12	5	1
(2)	4	4	2	2	9	2	6	4	—
(3)	2	—	3	4	2	5	3	—	—

Although there were no silver coins in this find, the Anglo-Saxon laws of the period imposed fines that imply payment in

---

* *Num. Chron.*, Jan. 1844, p. 171.
† *Anglo-Saxon Gold Coinage in the light of the Crondall Hoard*, by C. H. V. Sutherland, 1948.

terms of a silver currency, and certainly this metal predominated when the Carolingian replaced the Merovingian period on the Continent.

Henry III's pure gold penny of 1257 was the first post-Conquest gold coin; it weighed 2 pennyweights tower and its official value was 20 pence in terms of the current sterling penny. If this weighed a pennyweight, the (pure gold/sterling) value ratio was 10, and that this proved too low is evident from the fact that the denomination of the gold piece was raised in 1265 to 2 shillings, making the (gold/sterling) value ratio 12. As sterling contains 7·5 per cent alloy, this ratio was 11·1 for pure metals. At its original value the gold penny offered 2·25 gt. of pure gold for a sterling penny; its revaluation reduced the pennyworth of gold to 1·875 gt.

The introduction of this gold penny in England may have been inspired by the good reception accorded on the Continent to a gold coin, worth about 3 shillings, issued at Florence in 1252: in any case, the next English gold was struck by Edward III in 1344 from dies designed by Florentine craftsmen, and the coin was called a florin. Made of 'standard gold' containing (1/8) carat of alloy in 24 carats it was much over-valued at 6 shillings and was replaced within the year by the 80-penny noble, which gave the half-mark a place in the coinage. This change reduced the (G/S) for pure metals from 13·77 to 11·9. War and increasing trade raised the price of silver and correspondingly reduced the mass of the sterling penny to the point at which 300 were being shorn from the pound tower: by the end of this reign the (G/S) for pure metals was 11·14.

Sterling increased in value to half a crown an ounce (tower) under Henry IV; that is the mass of the penny fell to 15 gt. and the noble was reduced to 108 gt., which reduced the (G/S) to 10·3. This ratio remained when Edward IV in 1464 raised the value (rating) of the noble to 100 pence in terms of a penny that had fallen to 12 gt. At 100 pence, however, the noble ceased to have the significance of half-mark, and a new 80-penny piece called angel was coined in the following year at a weight that restored the (G/S) to 11·14. As this coin weighed 80 gt., the price of standard gold became a penny a grain (troy). Similarly, the

Some of the principal gold coins

(The dates relate to the gold; the silver was current some time during the same period.)

P = Pure gold = (24/24).   S = Standard gold = (191/192).
C = Crown gold = (22/24).   Sterling silver = (111/120).

Number of coins per *lb. tower   lb. troy*
　　　Gold　　　　　　　N1　　　　N2
Sterling pennies　　　　n1　　　　n2
V = Value of gold coin in shillings and pence.
(G/S) = Pure (gold/silver) value ratio.

Gold coins

Reign		Coin		N1	V	n1	(G/S)
Hy. III	(1257)	Penny	P	120	1/8	240	9·25
,,	(1265)	,,	,,	120	2/–	240	11·1
Ed. III	(1344)	Florin	S	50	6/–	243	13·77
,,	,,	Noble	,,	39	6/8	243	11·9
,,	(1346)	,,	,,	42	,,	270	11·57
,,	(1351)	,,	,,	45	,,	300	11·14
Hy. IV	(1412)	,,	,,	50	,,	360	10·3
Ed. IV	(1464)	,,	,,	50	8/4	450	,,
,,	(1465)	Angel	,,	67·5	6/8	450	11·14
,,	,,	Ryal	,,	45	10/–	450	,,
Hy. VII	(1489)	Sov.	,,	22·5	20/–	450	,,

				N2		n2	
Hy. VIII	(1527)	Sov.	S	24	22/6	576	10·45
,,	,,	George	,,	81	6/8	576	,,
,,	,,	Crown	C	100·5	5/–	576	10·56
,,	(1544)	Sov. 23 ct.		28·8	20/–		(debased silver)
,,	(1545)	Sov.	C	30	,,		,,
Ed. VI	(1551)	Angel	S	72	10/–		,,
,,	,,	Sov.	,,	24	30/–		,,
,,	,,	,,	C	33	20/–	720	11·1
Eliz.	(1601)	,,	,,	33·5	,,	744	10·89
James I	(1604)	Ryal	S	27	30/–	744	12·15
,,	,,	Unite	C	37·2	20/–	744	12·11
Chas. II	(1662)	Broad	,,	41	,,	744	13·34
,,	(1670)	Guinea	,,	44·5	,,	744	14·46
Geo. I	(1718)	,,	,,	44·5	21/–	744	15·2
Geo. III	(1816)	Sov.	,,	46·725	20/–	*	14·26

---

* The silver was currently coined at 5s. 6d. per oz. troy (= 792d./lb. troy) but the sterling itself had retired from business life early in the reign of George II: it is still alive as the penny in Maundy money and this is legal tender.   Copper pennies weighing 1 oz. averdepois were struck in 1797 and gave place in 1860 to the modern bronze in which 3 new pennies weigh 1 oz. and the diameter of the halfpenny is 1 in.

10-shilling ryal (also called rose noble) weighed 120 gt. and the 20-shilling sovereign struck by Henry VII in 1489 weighed 240 gt.

Half way through the reign of Henry VIII the higher price of silver brought the mass of the penny down to 10 gt., which resulted in the sovereign and angel being proclaimed worth 22s. 6d. and 7s. 6d. respectively. Wolsey, to whom Henry had referred the money problem in 1526, had revalued the sovereign tentatively at 22s. and had introduced a new gold coin at 4s. 6d., based on the French écu; it is known as the crown of the single rose to distinguish it from the crown of the double rose issued at 5 shillings to reinstate the half-ryal. The half-mark was reinstated at the same time and called the George (noble) in reference to the St. George and the Dragon portrayed on its reverse. This noble was in standard gold, but the new crown was 22 ct., and gave its name to the 'crown gold' that subsequently displaced 'standard gold' from the coinage. In 1544 the silver coinage was debased as a source of profit for the king; the mass of the penny remained 10 gt. but its silver content was so reduced ultimately that passable imitations were forged in copper. Just prior to the debasement the (G/S) was 10·5.

When the penny emerged as sterling in the sixth year of Edward VI's reign it weighed only 8 gt. In 1551 it was current with a 20-shilling sovereign in crown gold that restored the (G/S) to 11·1 and there it remained until, near the end of Elizabeth's reign, the penny and the sovereign were both reduced in mass and the (G/S) became 10·89 temporarily before rising sharply to 12·1 under James I and to 14·46 in 1670 under Charles II. It was during this reign that a revolutionary change took place in the appearance of the coinage; in 1662 the geometric perfection achieved by the mechanical mill relegated to the realm of antiquity those pleasing irregularities of outline that characterize coinage struck under the moneyer's hammer. It was a foretaste of a mechanized world in which gold was to oust silver as the standard of monetary value. Also in this reign the 20-shilling piece acquired the name guinea in commemoration of a new source of its gold.

Under William and Mary the value of the guinea increased to

30s. in 1694 but it was reduced by statute to 21s. 6d. in 1698 and in 1717 Sir Isaac Newton (then Master of the Mint) estimated it to be worth 20s. 8d. It was in this reign (George I) that it acquired the 21s. rating that it retained until its final issue under George III. For a century the 21s. guinea was the principal gold coin but in 1816 another (and final?) reinstatement of the 20-shilling sovereign was made with a lighter coin designed by Pistrucci. Its weight was such that 1869 of them could be coined from 40 lb. troy of 22 ct. gold (commonly now called standard gold), which is equivalent to £4 4s. 11·5d. per troy ounce for pure gold. Also in 1816 the status of the sterling coinage as legal tender was restricted to maximum amounts of 40s., and gold thus became the only unlimited standard of monetary value. The discovery of new mines enabled an adequate supply of bullion to be available at the statutory price for nearly a century, but with the outbreak of the war in 1914 gold currency was soon replaced by paper.

XIV (9) CONSTRUCTION OF AN APPROXIMATE RADIAN

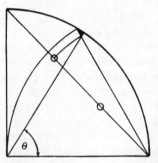

By chance I observed that the angle at the centre of a circle of royal cubit radius is rather close to a radian when it is subtended by a length equal to the Assyrian cubit marked off from one end of the sextant chord.

As the Assyrian cubit is two-thirds of the quadrant chord in this circle, the proposition can be generalized by saying that two-thirds of the quadrant chord of any circle, marked off on its sextant chord, subtends at the centre of that circle an approximate radian θ. The error appears to be less than 1 part in 300.

Suppose a radian to be formed by a radius vector moving from the boundary of a sextant; then the angular movement required is

$((\pi/3) - 1) = 0.0472$ radian, and a small triangle is formed in which one side is a short arc of the circumference. Its angles are not far from 30, 60, and 90 degrees; and, therefore, its hypotenuse is close to $(2/\sqrt{3})((\pi/3) - 1) = 0.0545R$. The amount of the sextant chord, therefore, that does subtend a radian is close to $(1 - 0.0545) = 0.9455R$; and this is less than 1 part in 300 in excess of two-thirds of the quadrant chord; that is $(2/3)\sqrt{2}R = 0.943R$.

A radian is the angle that subtends, at the centre of the circle, an arc equal in length to the radius. The circle measures 2π radians ($=360°$). In terms of the radian-arc, the quadrant-arc and quadrant-chord measure $(\pi/2)$ and $\sqrt{2}$ respectively.

XV

Bibliography

(1) Relating to ancient foreign weights and measures.
(2) Relating to the Pyramids.
(3) Statutes and other principal sources of information relating to the history of English metrology.
(3) Other books containing references to English metrology.

XV(1) Books Relating to Foreign Weights and Measures

Agricola.

De re metallica, by Georgius Agricola. Translated from the first Latin edition of 1556, by Herbert Clark Hoover and Lou Henry Hoover, 1912. This translation was made by Mr. and Mrs. Hoover when they lived in London (before 1914) at the Red House, Kensington, and had no premonition of the fate that would transfer them to the White House at Washington. The book ranks as a classic in the early bibliography of mining: its relevance to metrology lies in the author's references to the weights with which he was familiar in sixteenth-century Germany. These show the survival of the Roman 8-uncia mark called bes as the major unit of three systems; thus:

1. 'divided 12 times into units of 5 drachmae and 1 scruple each, which ordinary people call nummi; each of these we again divide into 24 units of 4 siliquae each, which the same ordinary people call grenlin.'

2. 'divided into 16 semunciae which are called loths, each of which is again divided into 18 units of 4 siliquae each called grenlin.'

3. or each semuncia 'is divided into 4 drachmae and each drachma into 4 pfennig.'

Siliqua Grenlin Pfennig Scruple Drachm Loth Nummi Bes

1152 = 288 = 256 = 192 = 64 = 16 = 12 = 1

This evidence of the survival of the bes draws attention to the (224/225) mass ratio of the double-bes (= 16 unciae = 6720 gt.) to 15 tower oz. = 6750 gt. = 25-shillingsweight tower: it is apparent that these standards were likely to be confused if they ever operated in the same market.   In England the tower dimark (= 16 tower oz.) and the 25-shillingsweight troy (= 15 troy oz.) were identical in mass.

Among Agricola's descriptions of mining practice is one disclosing the use of proportional weights on the 'assay ton' principle.   (See XIII.)   Also he explains how the fineness of gold or silver can be estimated from the colour of its scratch on a touchstone, and he illustrates the sets of 'needles' of known (graduated) composition used to produce the matching colour of identification.   Touchstone is a fine-grained variety of quartz or jasper (*O.E.D*) and Agricola says it must be 'thoroughly black and free from sulphur.'

Airy.
'Origin of the British Measures of Capacity, Weight, and Length.'   Wilfred Airy, in the *Proceedings of the Institution of Civil Engineers*, vol. clxxvii, part 3, p. 164, 1908–9.   Includes the masses of selected Egyptian weights in the British Museum.

Belaiew.
'On the Sumerian mina; its origin and probable value,' by N. T. Belaiew, *Newcomen Society*, vol. viii, 1927–8.
'Weights found at Susa.'   *Mémoires de la mission archéologique de Perse*, vol. xxv, p. 134, 1934.
An article (in Russian) on ancient measures, in which the author expresses the (Greek/Royal) cubit length ratio as (1/2) $\sqrt{\pi}$.   *Seminarium Kondakovianum* (1927).

Colebrooke.
Algebra with Arithmetic and Mensuration from the Sanskrit of Brahmagupta and Bhascara, translated by Henry Thomas Colebrooke, 1817.   This includes the 'Lilavati.'

Cornish.
Dictionary of Greek and Roman Antiquities, by F. W. Cornish, 1898.

Cumberland.

An essay towards the recovery of the Jewish measures and weights, by Richard Cumberland, 1685.

The author was Bishop of Peterborough; and his essay is dedicated to his friend Samuel Pepys, then President of the Royal Society.   In the preface he says:

> I believe this book will be the more welcome into your choice library because the subject of it is not any quarrelsome interest but the peaceable doctrine of measures and weights, which, in their general nature, are the common concern of all mankind; as being the necessary instrument of just dealing, and fair commerce between nations; which the Admiralty of England doth promote in times of peace as it secures our safety in times of war.

Although this commentary was not an official document, Cumberland's contribution became the generally recognized interpretation of biblical weights and measures.   He adopted the Egyptian derah, measured by Greaves at Cairo, as the cubit.   In my analysis this cubit is called Talmudist and rated (3/2) remen = 30 digits = 21·87 in. (Greaves 21·888 in.): the sacred cubit was longer.

Daremberg and Saglio.

Dictionnaire des Antiquités Grecques et Romaines d'après les textes et les monuments, C. H. Daremberg and Edm. Saglio, 1875.

Epiphanius.

Treatise on Weights and Measures, by Bishop Epiphanius (A.D. 392).   Translated, from the Syriac version in the British Museum, by James Elmer Dean, 1935.   No. 11 in the series 'Studies in Ancient Oriental Civilization.'   (Oriental Institute of the University of Chicago.)

Epiphanius was Bishop of Constantia in Cyprus, and already seventy-seven when he wrote this treatise from material compiled some years previously, when the Emperor Valentinian II summoned him to instruct a Persian priest who was

> eager to learn whatever is of value in the divine Scriptures.   He found weights and measures in the divine Scriptures; he sought information about them from Saint Epiphanius, the bishop.

Epiphanius had some reputation as a linguist, knowing Hebrew, Syriac, Egyptian, Greek, and Latin: the oldest surviving fragments of his *Weights and Measures* are written in Greek, but the only complete versions are the two Syriac translations in the British Museum; the older of these was written between A.D. 648 and 659. Epiphanius died in A.D. 403 at the age of eighty-eight. He is commemorated by one of the many medallion portraits that form the great coloured frieze lately revealed by alterations at the Bodleian Library.

Evans.

The Palace of Minos at Knossos, by Sir Arthur Evans. 'Minoan Weights': *Corolla Numismatica.*

Folkes.

'An account of the Standard Measures preserved in the Capitol at Rome,' by Martin Folkes. *Phil. Trans.*, No. 442, p. 262, 1736.

Gardiner.

Egyptian Grammar (Lesson on Weights and Measures), by A. H. Gardiner.

Greaves.

Miscellaneous Works of Mr. John Greaves, Professor of Astronomy in the University of Oxford, published by Thomas Birch, 1737.

These two volumes contain reprints of *Pyramidographia*, 1646, the 'Discourse on the Roman Foot and Denarius,' and Newton's 'Dissertation on the Cubit': they are prefaced by a biography.

Greaves was the first English authority on Ancient Metrology: his measurements in Egypt and Rome formed the factual basis of dissertations by Newton (whose unpublished text, translated from the Latin, is included by Birch); by Cumberland, who prepared an interpretation of biblical metrology, and by several other men of note then interested in this subject. He was Gresham Professor of Geometry in London at the time he made his measurements in Rome and Egypt, but in 1640 he went to Oxford as Savilian Professor of Astronomy. This was two years before the outbreak of the Civil War; the discordant times that ensued brought him much trouble.

In the *D.N.B.* a posthumous publication attributed to Greaves is entitled '*The Origin and Antiquity of our British Weights and Measures* discovered by their near agreement with such standards as are now found in one of the Egyptian Pyramids. Together with the explanation of divers lines therein heretofore measur'd by Mr. John Greaves. Astronomy Professor at Oxford.' 2nd ed. 1745. The title-page is set out to make Greaves appear as the author, but the book is a collection of unsigned letters to an unnamed correspondent: Greaves's name is mentioned frequently, also events dated long after his death. An item of interest is the trial measurement in 1696 of the Exchequer copy of the Winchester bushel.

Greek Coins.
Les Poulains de Corinthe, by O. E. Ravel, 1936.
Greek Coins by C. Seltman, 1933, also *Athens, its History and Coinage*, 1934.
Greek Coinage, by J. G. Milne (1931), also articles on Solon's reforms, in the *Journal of Hellenic Studies*, vol. l, p. 179 (1930), and in the *Classical Review*, March 1943.

Griffith.
'Notes on Egyptian Weights and Measures,' by F. L. Griffith. *Proceedings of the Society of Biblical Archaeology*, vol. xiv, p. 403, 1891–2; vol. xv, p. 304, 1892–3. This is an important discussion of the metrological evidence, by a former professor of Egyptology in the University of Oxford.

Hallock and Wade.
Outlines of the Evolution of Weights and Measures and the Metric System, by W. Hallock and H. T. Wade, 1906.

Hussey.
An Essay on the Ancient Weights and Money and the Roman and Greek Liquid Measures, with an Appendix on the Roman and Greek Foot, by Robert Hussey, 1836. A study of the metrological content of the classical texts and of the relevant physical evidence, by a former student of Christ Church who became Regius Professor of Ecclesiastical History. Of the Vespasian congius he reports that 'before 1721 it had been removed to Dresden where it remained long unnoticed until (in 1824) Dr.

Hase recognized it and published (its water-weight as 63460 French grains) in the Memoirs of the Academy of Berlin (Histor. Philog. Klasse, p. 149). Dr. Hase's weight agrees very nearly with that which Auzout obtained. . . .'

Hyde.

Epistola de mensuris et ponderibus Serum seu Sinensium, by Thomas Hyde, 1688. The author was Bodley's Librarian, and he obtained from a Chinese undergraduate the information about Chinese metrology.

I.C.T.

The International Critical Tables of Numerical Data, prepared under the auspices of the International Research Council and National Academy of Sciences by the Research Council of the U.S.A. Volume i contains a tabulated summary of the 'best studied' systems of antiquity, but there is neither comment nor reference to the sources of information.

Jomard.

Mémoire sur le système métrique des anciens Égyptiens contenant des recherches sur leurs connaissances géométriques et sur les mesures des autres peuples de l'antiquité, by E. Jomard, 1817. The author was in the team of scientists with Napoleon's army during the invasion of Egypt. He noticed the geodetic appearance of the Greek linear scale.

Description d'un étalon métrique, 1822. This is the earliest description of the Harmhab cubit-rod.

Langdon.

'Note on Dudu's mina.' S. Langdon, in the *Journal of the Royal Asiatic Society,* 1921, p. 575.

Lepsius.

Die Alt-ägyptische Elle, by R. Lepsius, 1865. See VI(6).

Lucas and Rowe.

'Report on certain measurements made in the Cairo Museum,' by A. Lucas and Alan Rowe, published in the *Annales du Service Antiquités de l'Égypte,* t. xl., 1939. Includes details of the two medicine cups.

Manu.

The Laws of Manu, *c.* 500 B.C. (Buhler's translation, 1886). Contains a statement relating to weights.

Michaelis.

'The Metrological Relief at Oxford,' by Ad. Michaelis. *Journal of Hellenic Studies*, vol. iv, p. 335, 1883.

This is a description of one of the Arundel marbles presented to the Ashmolean Museum in 1755 by the Dowager Countess of Pomfret.   See IX(7).

Mohenjo-daro and Harappa (excavations).

Mohenjo-daro and the Indus Civilization.   Official report of the Government of India's excavations between the years 1922 and 1927.   Edited by Sir John Marshall, 1931.

Further Excavations at Mohenjo-daro from 1927 to 1931.   Official account edited by Dr. E. J. H. Mackay, 1938.

Excavations at Harappa.   Official account, edited by Madha Sarup Vats, 1940.

The chapters on weights in all these reports are by A. S. Hemmy.   He recognized 27·2 gm. as the mass standard of the numerous small cubes of chert.

Moreland.

'Notes on the Indian Maunds,' by W. H. Moreland. *Indian Antiquary*, vols. lx and lxi, 1931–2.

Neugebauer and Sachs.

Mathematical Cuneiform Texts.   Edited and translated by O. Neugebauer and A. Sachs.

Newton.

A Dissertation on the Sacred Cubit of the Jews.   Translated from the Latin of Sir Isaac Newton.   Published, with Greaves's works, by T. Birch, 1737.

Oppert.

'L'Étalon des Mesures Assyriennes fixé par les textes cunéiformes,' by M. J. Oppert, *Journal Asiatique*, Aug.–Sept. 1872 and Oct.–Nov. 1874.   Includes a summary of measurements made at Babylon in 1853.

Peet.
The Rhind Mathematical Papyrus. Translation and commentary by T. E. Peet, 1923.
The introduction reads:

Rules for enquiring into nature and for knowing all that exists, every mystery . . . every secret. Behold this roll was copied in year 33 month 4 of the inundation season. . . . Under the Majesty of the King of upper and lower Egypt Aauserre, endowed with life, according to a writing of antiquity made in the time of the King of upper and lower Egypt, Nemare. It was the scribe Ahmose who made this copy.

Penrose.
An investigation of the Principles of Athenian Architecture, by Francis Cranmer Penrose, 1888. Contains measurements of the Parthenon.

Petrie.
Inductive Metrology, 1877. *Ancient Weights and Measures,* 1926. *Wisdom of the Egyptians,* 1940. 'Ancient Weights and Measures,' by W. M. Flinders Petrie, in the 14th edition of the *Encyclopaedia Britannica.*

Sarton.
Introduction to the History of Science, by George Sarton. Vol. i. 1927.

It might be worth while to put together the data (relating to ancient weights and measures) collected from the archaeological monuments but that would be a considerable task and conclusions should be suspended until it is accomplished.

Sarzec E. de.
Découvertes en Chaldée, by Ernest de Sarzec, 1884-1912. Contains the report of the original excavations at Lagash; illustrations of Entemena's silver vase, and of the linear scales on Gudea's statues, are included.

Silberrad.
Report of the (Indian) Weights and Measures Committee. (C. A. Silberrad. Chairman.) 1913-14.

Stein.

Serindia. A detailed report of explorations in Central Asia and Westernmost China, carried out and described under the orders of H.M. Indian Government, by Sir Aurel Stein, 1921.

This was the author's second expedition (1906–8), made 'in the hope of recovering from ruins long ago abandoned to the desert more relics of that ancient civilization which the joint influence of Buddhist-India, China, and the Hellenized Near East had fostered in the scattered oases of those Central-Asian passage lands.' The region covered was roughly that between latitudes 33–45 N., longitudes 75–102 E. The index to metrological material being incomplete, these page numbers are given for reference. Measures: 373, 374, 378, 433, 434, 559, 649, 660, 668, 671, 672, 701, 734, 773, 757, 1464. Weights: 121, 316, 1465.

Stuart.

The Antiquities of Athens, by James Stuart and Nicholas Revett, 1762–1816.

On the publication of the first volume the knowledge of Grecian art burst upon the public in all its splendour: its author acquired the surname Athenian, *par excellence.* (Editor, vol. iv, 1816.)

It is the second volume, published posthumously, that contains measurements of the Parthenon. (For measurements made a century later see Penrose.)

Thureau-Dangin.

Textes Mathématiques Babyloniens, by F. Thureau-Dangin, 1938.

Walston.

The Argive Heraeum, by Sir Charles Walston, 1902.

.. The excavations on this site were carried out by the American School of Classical Studies at Athens, under the author's direction, from 1892 to 1895.

The Argive Heraeum was the foremost sanctuary in the Peloponnesus; serving Mycenae in this capacity before Argos rose to pre-eminence. It housed the statue of Hera, sculptured in ivory and gold by Polycleitus; and here Pheidon deposited a bundle of iron currency spits to commemorate his introduction of the first silver coinage. The Heraeum was situated three miles south-east of Mycenae, five miles north-east of Argos, and about six miles north of Tiryns.

Wang Kuo-wei.
Celebrated Chinese scholar (1877–1927): his manuscript of a lecture on 'Chinese foot-measures of the past nineteen centuries' was translated by Arthur W. Hummel, and published in the *Journal of the Royal Asiatic Society* (North China Branch), 1928.

Watson.
'Babylonian Measures of Length,' by Col. Sir C. M. Watson. *Proceedings of the Society of Biblical Archaeology*, vol. xxxvii, p. 60, 1915. The author comments on the geodetic appearance of the Greek linear scale.

XV(2) BIBLIOGRAPHY RELATING TO THE PYRAMIDS

Cole.
'Determination of the exact Size and Orientation of the Great Pyramid at Gizeh,' by J. H. Cole. *Survey of Egypt.* Paper No. 39, 1925.

Drioton and Lauer.
Sakkorah. The Monuments of Zoser, by E. Drioton and J. P. Lauer, 1939.

Edwards.
The Pyramids of Egypt, by I. E. S. Edwards, 1947.

Greaves.
Pyramidographia (1646) is a description 'showing in what manner I found the pyramids in the years 1638 and 1639 or 1048 of the Hegira. For I twice went to Grand Cairo . . . carrying with me a radius of 10 feet most accurately divided (into 20000 parts) besides some other instruments for the fuller discovery of the truth.'
Reprinted in 1737 by Thomas Birch, under the title *Miscellaneous Works of Mr. John Greaves.*

Herschel.

'Observations on the Entrance Passage to the Great Pyramid,' by Sir John Herschel.   This is a note written for Col. Howard Vyse and published in his *Pyramids of Gizeh*.

Herodotus.

II. 124–6.   Description of the Great Pyramid.

Lauer.

La Pyramide à degrés, by Jean-Philippe Lauer, 1936.   This is a description of the earliest Pyramid.   *Le Problème des Pyramides d'Égypte*, 1948.   Contains a chapter on 'Connaissances scientifiques—La Géométrie des Pyramides.'

Perring.

'Survey of certain Pyramids in Egypt,' by J. S. Perring.   Vol. iii of the *Operations carried on at the Pyramids of Gizeh in* 1837, by Col. Howard Vyse.

Petrie.

The Pyramids of Gizeh, survey and description, 1883.

The Royal Society made a grant in aid of the publication of this investigation.   Petrie's measurements of the base of the Great Pyramid were confirmed in 1925 by Cole measuring officially for the Survey of Egypt.

Piazzi Smyth.

Life and Work at the Great Pyramid, 1867.   *Our Inheritance in the Great Pyramid*, 4th edition, 1880, by Piazzi Smyth (Astronomer Royal for Scotland).

Inspired by Taylor (to whom the book is dedicated) the author decided to make his own measurements, and these are a contribution to the factual evidence; but while he was in Egypt a letter from a young stranger started a supplementary idea in his mind and led him to give prophetic meaning to some of his measurements.

Taylor.

The Great Pyramid.   Why it was built and who built it, by John Taylor, 1859.

This book contains the earliest reference to the π function in the angle of the casing stones; but, seeing no geometry in the other pyramids measured by Vyse and Perring (1837), Taylor allowed himself to believe that the Great Pyramid must have been designed and built under divine inspiration, perhaps by Noah but certainly by no Egyptian.

Piazzi Smyth and others who fell under the emotional influence of Taylor's ideas, caused the Great Pyramid evidence relating to Egyptian geometry and metrology to become suspect. Taylor's other information relating to ancient metrology shows his extensive reading.

Vyse.
Operations carried on at the Pyramids of Gizeh in 1837, by Col. Howard Vyse, 1840.

XV(3) Statutes and other Principal Sources of Information Relating to the History of English Metrology

Ancient Laws and Institutes of England.
Comprising laws enacted under the Anglo-Saxon Kings from Aethelbirht to Cnut, the laws called Edward the Confessor's, the laws of William the Conqueror, and those ascribed to Henry I, etc. Printed by Command of His Late Majesty King William IV under the direction of the Commissioner of the Public Records of the kingdom. MDCCCXL.

Statutes of the Realm.
Printed by Command of His Late Majesty King George the Third in pursuance of an address of the House of Commons of Great Britain. From the original records and authentic manuscripts. 1810.

The Statutes at Large.
The Statutes at Large from Magna Charta to the end of the last Parliament 1761, by Owen Ruffhead (1763).

Irish Statutes.
The Statutes at Large passed in the Parliament held in Ireland from the third year of Edward the Second A.D. 1310 to the

twenty-sixth year of George the Third A.D. 1786.   Published by Authority.

Studies on Anglo-Saxon Institutions, by H. Munro Chadwick, 1905.

The Laws of the Earliest English Kings.   Edited and Translated by F. L. Attenborough, 1922.

The Laws of the Kings of England from Edmund to Henry I. Edited and translated by Agnes Jane Robertson, 1925.

Select English Historical Documents of the Ninth and Tenth Centuries.   Edited by F. E. Harmer, 1914.

Excise to Treasury Series.
Bound volumes (in the Library of H.M. Customs and Excise) of manuscript copies of the correspondence between the Commissioners of Excise and Lords of the Treasury.   Vol. iii contains the letter of 1688 relating to the Excise and Guildhall wine gallons: also that of 1707 relating to the Scots gallon.

Prior to the appointment of the Commissioners, the Excise was collected in the Counties (as the Customs were collected in the Ports) by persons or syndicates to whom the revenues were farmed: the papers of this earlier period are not in the archives.

Collection of Reports from Committees of the House of Commons.
This collection, 'now comprised into 15 volumes, was begun in 1773 by collecting into 4 volumes such reports as had been printed in a detached shape but were not inserted in the Journals of the House.'   The selection of reports in continuation of these 4 volumes, and constituting the bulk of this collection was made in 1803.

Vol. ii, p. 411.   1758 May 26 Weights and Measures
,,   ,, p. 453.   1759 April 11   ,,   ,,   ,,

Journals of the House of Commons.
1759 and 1760.   Reference to Lord Caryfort's Bill to give effect to the recommendations of the reports of 1758 and 1759.

Parliamentary Papers.
1819.   First Report of the Commissioners appointed to consider the subject of Weights and Measures.

Seventh Annual Report of the Warden of the Standards, 1873.
This report includes a historical review and abstracts from the Harleian MSS.

Philosophical Transactions of the Royal Society of London.
1856, p. 753. 'On the Construction of the new Imperial Standard Pound,' by W. H. Miller.

XV(4) OTHER BOOKS CONTAINING REFERENCES TO ENGLISH METROLOGY

Chaney.
Our Weights and Measures. A practical treatise on the Standard Weights and Measures in use in the British Empire. With some account of the Metric System, by H. J. Chaney, 1897.

Chisholm.
On the Science of Weighing and Measuring and Standards of Measure and Weight, by H. W. Chisholm, 1877.
The author was Warden of the Standards.

Coins.
Anglo-Saxon Gold Coinage in the light of the Crondall Hoard, by C. H. V. Sutherland, 1948.
Annals of the Coinage of Great Britain, by the Rev. Roger Ruding, 3rd ed., 1840.
English Coins from the Seventh Century to the Present Day, by G. C. Brooke, 1932.

Fleta.
'The Book called Fleta (a tract on Weights and Measures) which is thought to have been wrote in the reign of Edward I.' (Report of the 1758 Committee.)

Fleetwood.
Chronicon Preciosum, 1707.

Hopton.
'A Concordancy of Years. Newly composed and digested by Arthur Hopton, gentleman, 1612.'
Contains a chapter on weights and measures in which the averdepois pound is supposed to be 16 oz. troy.

Kelly. Metrology; or an Exposition of Weights and Measures, chiefly those of Great Britain and France, by P. Kelly, 1816.

The Universal Cambist and Commercial Investigator, by P. Kelly, 2nd Edition 1821. Supplement 1824.

Comprehensive summary of foreign weights, measures, money, and exchange; incorporating information obtained from H.M. Consuls in all parts of the world. They sent this in response to a circular dispatch from the Foreign Secretary, who acted on a decision of the Privy Council. The enterprise, however, was initiated by Kelly in a letter to the Board of Trade (1818) in which he pointed out the erroneous state of existing tables, and proposed a plan for their correction.

Kingdom.

An attempt to explain how the King's weigh-house and beams within the City of London came into the charge of the Worshipful Company of Grocers, by J. A. Kingdom, 1904.

McCaw.

Articles in the *Empire Survey Review,* by G. T. McCaw.

'The African Foot,' April 1939, p. 98. Explains the problems caused by the introduction of the International Prototype Metre.

'Linear Units Old and New,' Oct. 1939, p. 236; 'The Competitive Foot,' Oct. 1939, p. 255. Discusses the feet of 12 and 13·2 in.

Mile and League.

Article on 'Mile.' *Penny Cyclopaedia,* 1833–58.

'The Old English Mile,' by W. Flinders Petrie. *Proceedings of the Royal Society of Edinburgh,* vol. xii, p, 254, 1883.

England under the Norman Occupation, by James F. Morgan, 1858.

Geographical By-ways, by Sir Charles Arden-Close, 1947.

Miller.

Speeches in the House of Commons upon the equalization of the Weights and Measures of Great Britain, by Sir John Riggs Miller, 1790.

Contains the author's correspondence with Talleyrand, and an account of their joint endeavour to persuade Parliament to co-operate with France in the development of the metric system.

Miners' and Brenners' Dish.

Journal of the Derbyshire Archaeological Society, 1937 and 1938. Contribution by T. L. Tudor.

Mining Magazine (Aug., Sept., Oct., 1932). 'Derbyshire Mining,' by Leslie B. Williams.

Reprinted Glossaries. Series B. Published for the English Dialect Society. Manlove's *Rhymed Chronicle,* 1653.

Petrie. See XV(1).

Inductive Metrology contains evidence of the ancient use in England of the foot of 13·2 in.

Salzman.

English Trade in the Middle Ages, by L. F. Salzman.

Seebohm.

Customary Acres and their Historical Importance, by F. Seebohm, 1914.

Shuckburgh-Evelyn.

'An Account of some endeavours to ascertain a standard of weight and measure,' by Sir George Shuckburgh-Evelyn. *Phil. Trans.,* 1798, pt. 1, p. 133.

> I had, so early as the year 1780, taken up the idea of an universal measure, from whence all the rest might be derived, by means of a pendulum.

Contains a table of prices of commodities from the time of William I, which is also published in Ruding's *Annals of the Coinage of Great Britain.*

Vetusta Monumenta (1746).

Plate 69. Illustration of the Exchequer table of weights and measures dated 'Anno 12 Henrici Septimi.'

Wilde.

Weights and Measures of the City of Winchester, by Edith E. Wilde, 1931.

Index

(5) GEODETIC AND GEOMETRIC

(7) BUILDINGS

(8) PLACES